Pain in Psychiatric Disorders

Modern Trends in Pharmacopsychiatry

Vol. 30

Series Editor

B.E. Leonard Galway

Pain in Psychiatric Disorders

Volume Editors

David P. Finn Galway
Brian E. Leonard Galway

11 figures and 12 tables, 2015

KARGER Basel · Freiburg · Paris · London · New York · Chennai · New Delhi · Bangkok · Beijing · Shanghai · Tokyo · Kuala Lumpur · Singapore · Sydney

Modern Trends in Pharmacopsychiatry
(Formerly published as 'Modern Problems in Pharmacopsychiatry')

Prof. David P. Finn
Prof. Brian E. Leonard
Department of Pharmacology
Centre for Pain Research and
Galway Neuroscience Centre
National University of Ireland
Galway (Ireland)

Library of Congress Cataloging-in-Publication Data

Pain in psychiatric disorders / volume editors, David P. Finn, Brian E. Leonard.
p. ; cm. -- (Modern trends in pharmacopsychiatry, ISSN 1662-2685 ; vol. 30)
Includes bibliographical references and index.
ISBN 978-3-318-05573-3 (hard cover : alk. paper) -- ISBN 978-3-318-05574-0 (electronic version)
I. Finn, David P., editor. II. Leonard, B. E., editor. III. Series: Modern trends in pharmacopsychiatry ; v. 30. 1662-2685
[DNLM: 1. Mental Disorders--epidemiology. 2. Pain--epidemiology. 3. Comorbidity. W1 MO168P v.30 2015 / WL 704]
RC455.2.E64
362.2'0422--dc23
2015027765

Bibliographic Indices. This publication is listed in bibliographic services, including Current Contents® and PubMed/MEDLINE.

Disclaimer. The statements, opinions and data contained in this publication are solely those of the individual authors and contributors and not of the publisher and the editor(s). The appearance of advertisements in the book is not a warranty, endorsement, or approval of the products or services advertised or of their effectiveness, quality or safety. The publisher and the editor(s) disclaim responsibility for any injury to persons or property resulting from any ideas, methods, instructions or products referred to in the content or advertisements.

Drug Dosage. The authors and the publisher have exerted every effort to ensure that drug selection and dosage set forth in this text are in accord with current recommendations and practice at the time of publication. However, in view of ongoing research, changes in government regulations, and the constant flow of information relating to drug therapy and drug reactions, the reader is urged to check the package insert for each drug for any change in indications and dosage and for added warnings and precautions. This is particularly important when the recommended agent is a new and/or infrequently employed drug.

www.karger.com
Printed in Switzerland on acid-free and non-aging paper (ISO 9706) by Kraft Druck GmbH, Ettlingen
ISSN 1662–2685
e-ISSN 1662–4505
ISBN 978–3–318–05573–3
e-ISBN 978–3–318–05574–0

KARGER 125 YEARS

Contents

Preface

Despite the widespread interest and concern over the nature and impact of pain on the well-being of patients, research into the causes and treatment of pain has mainly been confined to the provinces of neurology and medicine. And yet, it has been estimated that pain is co-morbid in approximately 50% of patients with major depression, a frequency not dissimilar to that reported for patients with anxiety disorders.

Against this background it is not surprising to find that there is a growing interest in pain and its connection to the pathophysiology of major psychiatric disorders. This is particularly relevant to the management and treatment of psychiatric disorders, as pain complicates the patient's response and recovery.

The purpose of this book is to present an overview of the complex, bidirectional interactions between pain and some major psychiatric disorders. To achieve this, the twelve chapters comprising this volume are divided into four major sections, commencing with the first chapter on animal models for the study of co-morbid pain and psychiatric disorders [1].

Because of the clinical importance of inflammation in the causes and psychological consequences of pain, the next section [2–4] summarises the role of inflammation that links pain to depression and other psychiatric disorders. The third section [5–7] deals with specific neuromodulatory systems including monoamine and amino acid neurotransmitters, endovanilloids/TRPV1 and neuroactive steroids, focusing on their importance both as key substrates driving pathophysiology and also as potential therapeutic targets in co-morbid pain and psychiatric disorders. The final five chapters [8–12] are primarily concerned with the clinical manifestation of pain in different psychiatric and psychological disorders.

The format of the chapters is similar to the four previous publications in the pharmacopsychiatry series. All the chapters are written by internationally acclaimed experts in pain research and ancillary areas. Every chapter contains key references to the latest research.

The editors are aware that there are many areas of psychiatry in which pain has been implicated but which have not been covered in this book. However, it is anticipated that this volume will provide a stimulus to others to extend the information on the impact of pain in these areas and thereby further open up the possibilities of research and treatment for the benefit of patients and their families.

David P. Finn, Galway
Brian E. Leonard, Galway

References

1 Leite-Almeida H, Pinto Ribeiro F, Almeida A: Animal models for the study of comorbid pain and psychiatric disorders; in Finn DP, Leonard BE (eds): Pain in Psychiatric Disorders. Mod Trends Pharmacopsychiatry. Basel, Karger, 2015, vol 30, pp 1–21.

2 Leonard BE: Pain, depression and inflammation: are interconnected causative factors involved?; in Finn DP, Leonard BE (eds): Pain in Psychiatric Disorders. Mod Trends Pharmacopsychiatry. Basel, Karger, 2015, vol 30, pp 22–35.

3 Burke NN, Finn DP, Roche M: Neuroinflammatory mechanisms linking pain and depression; in Finn DP, Leonard BE (eds): Pain in Psychiatric Disorders. Mod Trends Pharmacopsychiatry. Basel, Karger, 2015, vol 30, pp 36–50.

4 Lees JG, Fivelman B, Duffy SS, Makker P, Perera CJ, Moalem-Taylor G: Cytokines in neuropathic pain and associated depression; in Finn DP, Leonard BE (eds): Pain in Psychiatric Disorders. Mod Trends Pharmacopsychiatry. Basel, Karger, 2015, vol 30, pp 51–66.

5 Benson C, Mifflin K, Kerr B, Jesudasan SJB, Dursun S, Baker G: Biogenic amines and the amino acids GABA and glutamate: relationships with pain and depression; in Finn DP, Leonard BE (eds): Pain in Psychiatric Disorders. Mod Trends Pharmacopsychiatry. Basel, Karger, 2015, vol 30, pp 67–79.

6 Madasu MK, Roche M, Finn DP: Supraspinal transient receptor potential subfamily V member 1 in pain and psychiatric disorders; in Finn DP, Leonard BE (eds): Pain in Psychiatric Disorders. Mod Trends Pharmacopsychiatry. Basel, Karger, 2015, vol 30, pp 80–93.

7 Mifflin K, Benson C, Kerr B, Aricioglu F, Cetin M, Dursun S, Baker G: Involvement of neuroactive steroids in pain, depression and anxiety; in Finn DP, Leonard BE (eds): Pain in Psychiatric Disorders. Mod Trends Pharmacopsychiatry. Basel, Karger, 2015, vol 30, pp 94–102.

8 Felice VD, Moloney RD, Cryan JF, Dinan TG, O'Mahony SM: Visceral pain and psychiatric disorders; in Finn DP, Leonard BE (eds): Pain in Psychiatric Disorders. Mod Trends Pharmacopsychiatry. Basel, Karger, 2015, vol 30, pp 103–119.

9 Swiergiel AH, Juszczak GR, Stankiewicz A: Genetic and epigenetic mechanisms linking pain and psychiatric disorders; in Finn DP, Leonard BE (eds): Pain in Psychiatric Disorders. Mod Trends Pharmacopsychiatry. Basel, Karger, 2015, vol 30, pp 120–137.

10 Peters ML: Emotional and cognitive influences on pain experience; in Finn DP, Leonard BE (eds): Pain in Psychiatric Disorders. Mod Trends Pharmacopsychiatry. Basel, Karger, 2015, vol 30, pp 138–152.

11 Bandelow B: Generalized anxiety disorder and pain; in Finn DP, Leonard BE (eds): Pain in Psychiatric Disorders. Mod Trends Pharmacopsychiatry. Basel, Karger, 2015, vol 30, pp 153–165.

12 Schmahl C, Baumgärtner U: Pain in borderline personality disorder; in Finn DP, Leonard BE (eds): Pain in Psychiatric Disorders. Mod Trends Pharmacopsychiatry. Basel, Karger, 2015, vol 30, pp 166–175.

Finn DP, Leonard BE (eds): Pain in Psychiatric Disorders. Mod Trends Pharmacopsychiatry.
Basel, Karger, 2015, vol 30, pp 1–21 (DOI: 10.1159/000435929)

Animal Models for the Study of Comorbid Pain and Psychiatric Disorders

Hugo Leite-Almeida · Filipa Pinto-Ribeiro · Armando Almeida

Life and Health Sciences Research Institute (ICVS), School of Health Sciences (ECS), University of Minho, Braga, and ICVS/3B's – PT Government Associate Laboratory, Guimarães, Portugal

Abstract

Animal models of chronic pain have provided valuable information on the mechanisms of initiation and maintenance of the disease. Much of the research effort has targeted sensory abnormalities like hyperalgesia and allodynia. However, in the past 15 years a significant number of research groups have focused their attention on comorbid anxiety, depression and cognitive impairments that frequently emerge in chronic pain conditions. A myriad of paradigms have since then been introduced in the field to tackle multiple dimensions of rodents' behavior. Concerning emotional behavior, these include the elevated plus (and zero) maze and dark/light box for anxiety, the forced swimming and tail suspension tests for depression, and the spontaneous burrowing behavior for general well-being. Regarding the cognitive dimension, several water mazes (spatial-reference memory), attentional set-shifting test (attention and reversal learning), novel object recognition (memory), 5-choice serial reaction time task (sustained attention) and variable delay-to-signal task (impulsivity) are among the most commonly employed paradigms. The construct of some of these paradigms in the context of chronic pain will be reviewed in this chapter, with special emphasis on mood and cognitive alterations that are associated with the development of neuropathic and arthritic pain.

Chronic pain is a heavy burden for the individual and society. It affects 19% of the adult population in Europe and Canada [1, 2] and 30% in the USA [3], a tendency that clearly will rise with the ageing of the population in developed countries. It is estimated that total costs for chronic pain treatment reached USD 635 billion in 2010 in the USA, with additional healthcare costs due to pain ranging between USD 261 and 300 billion [4]. At the individual level, the patient with chronic pain is under continuous suffering, which deeply impacts his/her well-being, health status and social integration. This results in major comorbidities, namely psychiatric disorders like mood (depression and anxiety) and cognitive impairments [5, 6]. Importantly, a close reciprocal relationship occurs between chronic pain and anxiety, and chronic pain and depression [7], probably reflecting common mechanisms and neural pathways between these pathologies. For example, major depression affects 13–85% of patients with chronic pain (mean prevalence), whereas in patients with depression the prevalence of pain ranges between 15 and 100%, depending on the type of study [8].

Animal Measurement of Pain-Like Behaviors and Models of Chronic Pain

Although animal models in pain research have contributed to an enormous advance in unraveling pain mechanisms and treatments in humans, their use is still pertinent [9] and they have been at the center of debate in several studies [10, 11]. All types of models are known to have advantages as well as limitations. The first models of pain were based on spinal reflexes resulting from noxious stimulation applied by the experimenter (e.g. tail-flick, paw withdrawal) that can be performed in spinalized animals, and thus have a segmental level of neural mediation. Although this can be a reliable method to detect the analgesic properties of opioids, their simple evaluation is not compatible with the complex multidimensional pain experience in humans, which includes sensory, affective and cognitive components [12, 13].

Actually, the most used animal models of pain (table 1) are those mimicking, at least in part, some of the most common chronic pain pathologies, e.g. those concerning chronic inflammation of the joint(s) (arthritic pain) and those resulting from lesioning or disease of the nervous system (neuropathic pain) [10, 11], although there are putative models also for dysfunctional pain syndromes (e.g. fibromyalgia; [14]) (a detailed approach to these chronic pain models can be found below). Following the induction of chronic inflammatory or neuropathic models, which usually affects one of the animal's hind paws, classic signs of mechanical allodynia and hyperalgesia can then be measured using nociceptive tests like the von Frey filaments and the pressure application method (a variation of the Randall-Selitto test), respectively, whereas thermal (cold) allodynia and heat hyperalgesia can be measured using the acetone and Hargreaves withdrawal tests.

There are still problems in using animal models of chronic inflammation and experimental neuropathy due to poor construct and face and predictive validity, thus reducing the model's resemblance to human disease [10]. On the other hand, there are obvious contributions of animal models to the understanding of pain mechanisms and pain pathophysiology at the molecular, neuronal and system level, and to the identification of new targets and the development of new drugs [9]. Overall, the animal models of acute, inflammatory and neuropathic pain have the potential to predict the efficacy of drugs for specific analgesic mechanisms, at least for some clinically relevant compounds [15].

Table 1. Animal models of chronic pain

Pain	Models[1]	Reviews
Neuropathic (peripheral)		
Traumatic	Chronic sciatic constriction injury, spinal nerve ligation, spared nerve injury	[74]
Viral-related neuropathies	Varicella-zoster, HIV type 1 glycoprotein 120	
Ischemic	Chronic postischemia pain (complex regional pain syndrome type I	
Inflammatory	Sciatic nerve neuritis	
Malignant	Malignant sciatic neuropathy	[75]
Autoimmune	Autoimmune (antibody-mediated) neuropathy models	[76]
Metabolic	Diabetic sensory neuropathy models	[77]
Inherited neuropathies	Charcot-Marie-Tooth	[78]
Toxic exposure-related	Vincristine, cisplatin	[79]
Neuropathic (central)		
	Spinal cord injuries	[80]
	Thalamic syndrome (hemorrhagic)	
Arthritis		
Spontaneous	Age-associated[2], obesity models[2]	[60]
Surgical	Medial meniscectomy[2], temporomandibular joint anterior cruciate ligament transection[2], lateral collateral ligament transection[2]	[81, 82]
Chemical	Monoiodoacetate[2], kaolin and carrageenan[2], Freund's adjuvant[2]	[64, 83]
Enzyme/Autoimmune	Collagen-induced models[3], collagenase-induced models[2], antigen-induced arthritis[3]	[81, 84]
Genetically-modified	STR/ort mice[2], proteoglycan-induced models[3], K/B×N model[3]	[83, 84]
Dysfunctional		
Fibromyalgia	Repeated muscle insult, biogenic amine depletion, stress	[14]

Summary of chronic pain rodent models. [1] Representative examples. [2] Osteoarthritis experimental models; extensively reviewed by Handwerker and Arendt-Neilsen [37]. [3] Rheumatoid arthritis experimental models; extensively reviewed by Handwerker and Arendt-Neilsen [37].

Comorbid Psychiatric Disorders in Chronic Pain

It has long been observed that chronic pain patients present mood disturbances. While the sensory aspects of the disease have been extensively studied in the experimental models, it was only 15 years ago that the comorbid psychiatric disorders were addressed by basic researchers [16]. This section will focus on the experimental strategies used in this regard.

Evaluation of Emotional Behavior

The terms 'emotion' and 'emotional behavior' have been used in this field, rather indiscriminately to describe a number of states, namely anxiety, depression (anhedonia, despair), volition, fear and respective behavioral manifestations. We will not discuss its pertinence, but interested readers can find comprehensive reflections in a number of articles (e.g. [17]). In this section we will review the paradigms used to assess these phenotypes in rodents; we will not be exhaustive and the focus will be mainly on those that have been employed in the field of pain research. An extensive and authoritative review can be found in the work by Sousa et al. [18].

Elevated-Plus Maze, Zero Maze, Open-Field and Dark/Light Box

The elevated plus maze test is the gold-standard paradigm used for testing anxiety-like behavior in rodents [18]. It consists of an elevated plus-shaped maze with opposing open and closed arms and a center hub lying at the crossing between arms. The rationale of the test relies on the conflict between the rodent's innate drive to explore novel spaces and the aversion to open areas. Anxious profiles correlate with shorter periods of time spent in the open arms. Alternatively, the elevated zero maze can be used; in this case open and closed arms are sequentially organized in an annular fashion abolishing bias interpretations associated with the central hub [19]. The open field (OF) is probably one of the oldest behavioral paradigms [20]. Although structurally very different from the elevated plus maze/elevated zero maze, it also provides an effective tool to measure anxiety. In this case, a squared arena is used, but a similar conflict is established; more anxious animals prefer to explore the areas near the wall (thigmotaxis), avoiding the open spaces of the center of the arena. Additionally, the OF provides readouts of locomotor and exploratory behavior (distance and rearings). The dark/light box also uses the same rationale of the previous tests, but in this case the options lie between an enclosed (dark) and an open (light) compartment [21]; the OF arena can be easily adapted for this paradigm. An interesting and useful adaptation of the dark/light box used in the pain field is that of LaBuda and Fuchs [22]; in this case, animals receive a nociceptive stimulus (probing with von Frey filaments in the case of neuropathic animals) when they are in the less aversive compartment (dark) – the less aversive condition (light or pain) will determine the preference of the animal. These paradigms have been widely used in pain research (table 2).

Forced Swimming Test

Helplessness is an important manifestation of depression. The forced swimming test was developed in the mid-1970s as a screening test for putative antidepressant properties of drugs in rodents [23]. In this paradigm the animal faces an inescapable aversive scenario (a cylinder filled with water) leading to learned helplessness that is a hallmark of depressed individuals. The perseverance of the animal to escape the situation inversely correlates with depressive-like behavior. The forced swimming test has been extensively validated and is the most common test used to assess depressive-like behavior in rodents [18]. The tail suspension test can be alternatively used as a test for helplessness in mice [24]. In this case, suspending the animal by the tail for 5 min creates the inescapable situation; the amount of time the animal attempts to escape is the readout used.

Sucrose Preference Test

Anhedonia manifests in depression. Anhedonia can be assessed using the sucrose preference test. In this paradigm, animals are presented with two options: water and a sugared solution (e.g. 2%); anhedonia is reflected by a lack of preference for the

Table 2. Evaluation of anxiety- and depression-like behaviors in chronic pain models

Trait/paradigm	Model	Overall effect	Ref.
Anxiety			
EPM	Uric acid	Anxiety-like behavior 3 h after administration of uric acid	[85]
EPM; DB/LB	PSNL; CFA	Anxiety-like behavior in the EPM and DB/LB at day 28; EPM tested at 7 and 28 days; DB/LB tested at 7, 14, 21 and 28 days	[86, 87]
OF	VZV; PSNL; SNT	Positive correlation between mechanical hypersensitivity and anxiety in the VZV and SNT models at day 14	[41]
EPM	K/C	Anxiety-like behavior 5–6 h after induction	[66]
OF; DB/LB; EPM	SNL	Signs of anxiety-like behavior at postsurgery week 8 in all paradigms[1]	[46]
OF	PSNL; gp120	Signs of anxiety-like behavior at day 14; a second group was tested at day 7 but has not manifested any alteration	[88]
OF	ddC; ddC + gp120	Signs of anxiety-like behavior at day 14 (ddC + gp120) and 21 (ddC)	[89, 90]
EPM	SC	Anxiety-like behavior at day 30	[91]
DB/LB; EPM	PSNL	Anxiety-like behavior at postsurgery day 27 in both paradigms	[92]
EPM	CCI; PSNL	Anxiety-like behavior at day 28 (CCI); ≈40% less time in the OA PSNL compared to sham (no statistical significance)	[42]
EPM	Neonatal CFA	No sign of anxiety-like behavior in adults when pups were injected with CFA 15 days after birth, decreased anxiety-like behavior in adults animals when pups were injected with CFA 1 day after birth	[93]
EPM	SNI	Anxiety-like behavior 1 month after surgery in young and old animals	[51]
EPM	CCI	Anxiety-like behavior both in HAB and LAB animals at postsurgery day 36	[43]
EPM, EZM, TST	Obesity	Anxiety-like behavior in mice after 41 weeks of high-fat diet in all paradigms	[94]
EPM	SS	Anxiety-like behavior	[72]
OF; EZM; mb	SNI; CCI; CFA	Results vary according to the paradigm, sp/st and model 30 days after pain onset; generally, no alterations or decreased anxiety	[95]
DB/LB; mb; NSF	SC	Increased anxiety from the 4th up to the 8th postsurgery week	[45]
EPM/OF/social interaction	CCI	No alterations in anxiety-like behaviors in the EPM and social interaction tests at postsurgery days 14–21; decreased exploration of OF central area	[96]
OF	CFA	Anxiety-like behavior 7 and 14 days after induction	[97]
EPM	SNI	Anxiety-like behavior in left- but not right-sided SNI 1 month after surgery	[35]
DB/LB; NSF; contextual fear	SNI	SNI C57BL/6 mice presented and increased latency to leave the black box or to eat in a novelty context at postoperative day 12; contextual fear extinction was also impaired	[98]
EPM, OF, SIT, LB/DB	CFA	Anxiety-like behavior 28 days after induction although no difference in the LB/DB	[99]
Escape/avoidance; EZM	CCI	Anxiety-like behavior at day 28 in both paradigms	[100]
EPM, LB/DB	CFA	Anxiety-like behavior 1, 3 and 10 but not 7 days after induction in both paradigms	[101]
EPM, OF	CFA	Anxiety-like behavior 3 and 7 days after induction in both paradigms	[102]
OF, EPM	K/C	Anxiety-like behavior 4 weeks after induction in both paradigms	[64]
OF	SNI	Anxiety-like behavior at postsurgery day 21	[103]
EZM	CFA	Anxiety-like behavior 28 days after induction	[104]

Table 2. Continued

Trait/paradigm	Model	Overall effect	Ref.
EPM	CCI	Anxiety-like behavior at postsurgery day 25	[105]
OF; EZM	SC	Anxiety-like behavior 1 month after surgery	[106]
OF; PEAP	SCI	No signs of anxiety-like behavior at postsurgery days 28 and 42; increased place escape/avoidance	[107]
SI, OF	CFA	Anxiety-like behavior in all paradigms 2–3 weeks after induction	[108]
NSF	OXA	Anxiety-like behavior 7 days after the last OXA challenge	[109]
PEAP	CCI	Increased place escape/avoidance 7 days after surgery	[110]
EPM	CCI	Anxiety-like behavior at postsurgery days 7 and 21	[111]
EPM	PSNL	Anxiety-like behavior at postsurgery day 28 (but not 7, 14 or 21)	[112]
Depression			
FST	SNL	Depressive-like behavior at postsurgery week 8	[46]
FST	SNI	Depressive-like behavior at day 60	[113]
FST	CCI	Depressive-like behavior at postsurgery days 21–28	[114]
FST	SNI	No alterations in the latency to immobility index; increased immobility in the mid-aged SNI group in comparison with age-matched controls	[51]
FST	BAD	Depressive-like behavior	[71]
FST	SNI	Depressive-like behavior at postsurgery day 7	[115]
FST	SNL (L5)	Depressive-like behavior at postsurgery day 29	[116]
Splash test; FST	SC	Depressive-like behavior at postsurgery weeks 6–9	[45]
FST	CCI	Depressive-like behavior at postsurgery days 14–21	[117]
SPT, FST, TST	CFA	Depressive-like behavior 14 days after induction	[97]
FST	CCI	Depressive-like behavior at postsurgery day 28	[100]
SPT	PSNL	Depressive-like behavior at postsurgery day 10	[118]
SPT; FST	SNI	Depressive-like behavior at postsurgery day 14	[119]
TST, FST	CFA	Depressive-like behavior from day 7 to day 21 after induction in the TST, but only at 2 weeks in the FST	[120]
FST	SNI	Depressive-like behavior at postsurgery day 15	[121]
FST, SPT	K/C	Depressive -like behavior 4 weeks after induction in both paradigms	[64]
FST	CFA	Depressive-like behavior 28 days after induction	[104]
FST	CCI	Depressive-like behavior at postsurgery day 32	[105]
SPT	CCI	Anhedonia from the 4th to the 10th postsurgery week; normal values in the 1st to 3rd and 11th to 12th weeks	[122]
FST	SC	Depressive-like behavior 1 month after surgery	[106]
FST, TST	PSNL	Depressive-like behavior at postsurgery week 4	[123]
SPT	CFA	Depressive-like behavior 2–3 weeks after induction	[108]
TST	OXA	Depressive-like behavior 7 days after the last OXA challenge	[109]
FST	CCI	Depressive-like behavior at postsurgery days 7 and 21	[111]

Table 2. Continued

Trait/paradigm	Model	Overall effect	Ref.
TST; SPT	SCI	Depressive-like behaviors (helplessness and anhedonia) 2 months after surgery	[124]
FST, TST	CCI	Depressive-like behavior at postsurgery weeks 2, 4 and 6 (but not 1)	[125]
FST	CCI	Depressive-like behavior starting at week 0 (maintained up to week 3)	[126]

Summary of the literature concerning the effects of chronic pain in anxiety- and depression-like behavior in rodent models. Due to space constraints only positive results are shown. BAD = Biogenic amine depletion; CCI = chronic constriction injury; CFA = complete Freund's adjuvant; DB/LB = dark/light box; ddC = 2′-3′-dideoxycytidine (zalcitabine; antiretroviral agent); EPM = elevated plus maze; EZM = elevated zero maze; FST = forced swimming test; gp120 = immunodeficiency virus type I envelope glycoprotein 120; HAB = high anxiety-like behavior; K/C = kaolin/carrageenan; LAB = low anxiety-like behavior; mb = marble burying; n/a = not applicable (available); PEAP = place escape avoidance paradigm; NSF = novelty suppressed feeding test; OXA = oxaliplatin; PSNL = partial sciatic nerve ligation; SC sciatic nerve cuffing; SCI = spinal cord injury; SI = social interaction; SNI = spared nerve injury; SNL = spinal nerve ligation; SPT = sucrose/saccharine preference test; SS = sound stress; syst = systemic; SNT = spinal nerve transection; TST = tail suspension test; VZV = varicella zoster virus. [1] In a second experiment, a group of animals was tested at postsurgery days 2, 7, 15 and 30; depressive- and anxiety-like behaviors manifested at days 15 (FST) and 30 (OF, EPM), respectively.

latter. This test can be repeated multiple times and is therefore useful to assess the temporal evolution. Recently, our group has developed and validated the Sweet Drive Test for anhedonia which works on the same principles but allows faster readouts [25].

Spontaneous Burrowing Behavior

The spontaneous burrowing behavior test allows a readout of general well-being [26]. Animals are presented with a cylinder filled with gravel (or other substance) and the amount of displaced material during a defined period of time is quantified. Rodents have a natural drive to dig and therefore this is an ethologically relevant readout. On the downside, the paradigm is difficult to implement and a number of conditions including light, material and even pairing with 'natural diggers' might have to be tested beforehand. Importantly, for the field of pain a number of recent publications have reported that spontaneous burrowing behavior is reduced in chronic pain conditions and that it might be reversed with appropriated analgesic treatment [27, 28].

Evaluation of Cognitive Behavior

Cognition has been used as an umbrella word embracing different behavioral domains encompassing different subtypes of short- and long-term memory, decision-making, behavioral inhibition (impulsivity), attentional set-shifting, amongst others. From a functional perspective, each of these involves the recruitment of different networks; this diversity is matched by an elevated number of behavioral constructs that can be used to obtain behavioral readouts. The list presented in this section is limited and will mainly focus on those paradigms used in the pain field.

Water Mazes

Water mazes have been widely used to test learning and memory [29]. These are advantageous over other memory paradigms in many aspects, including that food or water deprivation is not required and that there is a near absence of intramaze cues (e.g. odors). On the other hand, unrelated strategies can be acquired, particularly egocentric orientation. In these paradigms, the animal, guided by cues placed in the walls of the testing room, has to reach a platform that is hidden slightly below the water level; the escape latencies or travelled distances are used as readouts. Working memory can be tested on a trial-to-trial basis by placing the platform in a new location every day (typically 4 trials/day are used). In the classical Morris water maze for long-term spatial referenced memory, the platform location is maintained in the same location every testing day (usually 4 days) and the daily evolution is studied. On the 5th day, the platform location can be shifted, allowing cognitive flexibility/reversal learning to be tested. Some precautions should be taken into account while interpreting water maze results, particularly the occurrence of unrelated strategies and possible motor deficits.

Attentional Set-Shifting Task

The attentional set-shifting task (ASST) is an analogue of the human Wisconsin card sorting task. In rodents it consists of the sequential acquisition of information leading to a reward and in the ability to reverse previous acquired rules [30]. The set of cues presented to the animal includes odors and textures simultaneously presented to the animal (compound discrimination). Following successful acquisition, reversal learning can be tested by changing the guiding cue to the second cue of the same dimension. After the introduction of new pairs of odors and textures, intradimensional shift or extradimensional shift can be tested if the relevant dimension is maintained or altered, respectively. In either case, the respective reversal tests can be performed. The ASST is advantageous over the Morris water maze/reversal as it allows discriminating prefrontal cortex subfunctions [30] and limits the use of alternative strategies, e.g. egocentric positioning. Food deprivation is required for a short period of time and should not constitute a major drawback.

5-Choice Serial Reaction Time Task

The rationale behind the 5-choice serial reaction time task (5-csrtt) follows that of the 5-csrtt developed by Leonard [31] to study sustained attention in humans. Animals are trained to poke the muzzle in a lighted aperture out of five possible choices in order to gain a sugared reward [32]. Overtime, the light signal duration is decreased, augmenting the attentional load of the test. In the 5-csrtt, nose pokes prior to the light signal (5-second delay) are registered as impulsive responses and determine the end of the trial (all lights off; no reward is delivered). Attentional deficits have been demonstrated in a model of chronic inflammatory pain [33]. Curiously, the number of premature responses has been shown to be decreased which most certainly reflects the increasing number of omissions (i.e. trials where no response is performed) over time.

Variable Delay-to-Signal Task

We recently designed and validated the delay-to-signal (VDS) task to measure impulsive behavior [34]. In this task, the same 5-csrtt apparatus is used, but the light is presented in a fixed location (middle aperture). After a training period with a fixed 3-second delay in which premature responses were signaled with the interruption of the trial (lights off; no reward), a single VDS session was performed. In this case, a block of 70 trials of large delays (6 and 12 s) was flanked by two blocks of 25 trials with a 3-second delay; in the case of this session, premature responses are allowed, i.e. they are registered but have no consequence for the trial sequence. The number of premature responses per minute of delay time is used as the readout. This task is advantageous over the 5-csrtt (or the delay discounting paradigm) in that animals learn it faster and more homogeneously, implying a shorter period of alimentary restriction. Additionally, the attentional component is minimized, facilitating the interpretation of the results. We have demonstrated that animals with a right-sided spared nerve injury (SNI) are more impulsive [35].

Animal Models of Chronic Pain

In this section we consider the division of types of pain as described by Costigan et al. [36]. Models of chronic pain are divided into neuropathic, inflammatory and dysfunctional types. Visceral and cancer pain models were not considered for the present review. A recent overall analysis of pain models and their relevance and applications can be found here [37].

Neuropathic Pain

Models

Neuropathic pain arises as a direct consequence of a lesion or disease affecting the somatosensory system [38]. The clinical conditions leading to neuropathic pain are heterogeneous implying that the underlying pathophysiological mechanisms need different modelling. This is reflected in the large number of animal models of chronic neuropathic pain available (table 1). Models were developed involving in most cases a traumatic lesion of a peripheral nerve, nerve roots or spinal cord caused by sectioning, cuffing, ligation, freezing, inflammation and/or ischemia [11, 39]. Additionally, drug toxicity and systemic metabolic disorders have also been used to trigger neuropathies. Some of the most commonly used models of chronic neuropathic pain are presented in table 1.

Impact on Emotional Behavior

In table 2, a summary of studies assessing anxiety- and depressive-like behaviors, anhedonia and other behavioral domains normally referred to as 'emotional' in models of chronic neuropathic pain chronic is given (see also the recent review by Yalcin et al. [40]). On one hand, the manifestation of emotional disturbances in chronic neuropathic pain conditions seems to be indisputable, either because of the large amount of observations pointing in that direction (table 2), the causality relations established by means of analgesic treatment and its ameliorating impact on the behavioral readouts, or by clinical reports in human patients (e.g. Andrews et al. [28]). On the other hand, it is virtually impossible to isolate the critical factors that determine the emergence of these manifestations. These could be of two levels: (1) related with the implementation of the behavioral paradigms (which are out of the scope of the present review), or (2) related to the neurobiology of the underlying pathophysiology of the disease. Concerning this latter point, the multiplicity of pain models used to study these phenomena is probably the greatest factor contributing to the discrepancies observed. Withdrawal thresholds and anxiety-like behaviors have been shown to correlate, suggesting that the type of model can be a determinant factor for anxiety manifestation [41, 42]. Conversely, trait anxiety was also shown to be a good predictor of pain magnitude [43] (similar observations were made in humans [44]). The temporal progression of the neuropathy was also shown to be a critical factor. Systematic analyses on the emergence of anxiety- and depressive-like phenotypes have revealed that these manifest at 4 and 8 weeks after pain initiation, respectively [45, 46]. Curiously, time-dependent maladaptive plasticity was observed in areas involved in emotion and cognition (e.g. prefrontal cortex) both in a rodent SNI model [47] and in human patients [48], suggesting a causal relationship. In the past years, our group has been tackling the effect of the neuropathic injury side. We demonstrated that left-sided pain was more anxiogenic than right-sided pain [35]. The underlying causes of such asymmetry are not clear, but might be related with the asymmetrical distribution of the functional load at the CNS. Indeed, Sullivan and Gratton [49] demonstrated that right- but not left-sided manipulations in the medial prefrontal cortex, depending on its nature, can be either anxiogenic or anxiolytic. Curiously, similar biases have also been observed in human patients [50]. Other factors might also be determinant for the emergence of pain-related comorbidities. We demonstrated, for instance, that chronic pain was more anxiogenic in older rats [51]; this is not a source of variability as virtually all studies were performed in young adult animals, but it should be taken into account when considering the translational potential of our models.

Impact on Cognitive Performance

In table 3, a summarized list of relevant articles testing cognitive function in chronic neuropathic pain models is given. The amount of studies produced in this field is smaller when compared with those investigating 'emotional' aspects (see 'Impact on Emotional Behavior'); such discrepancy is necessarily a reflex of the complexity behind many of the behavioral paradigms like the ASST, 5-csrtt or VDS (see 'Evaluation of Cognitive Behavior'), which require long experiments, appropriate instrumentation and know-how. Curiously, the number and diversity of the paradigms available, either subserving the same cognitive domain or different aspects of learn-

Table 3. Evaluation of cognitive performance in chronic pain models

Trait	Paradigm	Model	Overall effect	Ref.
Sustained attention	Delayed non-matching-to-position operant task	Polyarthritis (CFA)	Increased percentage of incorrect choices 1–2 months after pain onset	[127, 128]
Attention/memory	NOR	Colitis	Decreased attention toward a novel object up to 7 days after pain onset	[129]
Attention-demanding associative learning	Trace fear memory	Monoarthritis (CFA)	Impairments in trace fear memory	[130]
Attention	NOR	SNL/ Carrageenan	Decreased attention toward a novel object 4 h after carrageenan injection and 30 days after SNL	[131]
Memory	Water mazes	SNI	WM impairment in right- but not left-sided SNI 1 month after surgery (young subjects); no alterations in spatial reference MWM; reversal learning impairment in mid-aged subjects	[35, 51]
Sustained attention/ impulsivity	5-csrtt	Monoarthritis (CFA)	Decreased accuracy and increased omissions immediately after monoarthritis induction (sustained for 10 days); no difference in impulsive responses	[33]
Risk assessment	Gambling task	Monoarthritis (CFA)	Increased preference for risk choice 5, 21 and 56 days after pain onset	[132]
Spatial reference memory	MWM	SNL (L5)	Impaired learning performance at day 30	[116]
Risk assessment	Gambling task	K/C	Impairments in decision-making on the gambling task 5–6 h after induction	[133]
Attention/memory	NOR	PSNL	Decreased attention toward a novel object 7–9 days after PSNL	[134]
Working memory	8-arm; NOR	SNI	Increased number of WM errors and no differences in spatial reference memory errors in the 8-arm at postsurgery days 10–20 and 30–40; decreased short-term memory and no differences in long-term memory at postsurgery days 19 and 20, respectively, in the NOR	[56]
Social/working memory	Social recognition/ Y-maze	CCI	No deficits in the Y-maze; impairments at social recognition paradigm at postsurgery days 14–21	[96]
Attentional set-shifting and reversal learning	ASST	SNI	Impairments in reversal learning in right-sided SNI 1 month after surgery; extra- and intradimensional shift preserved	[35, 55]
Impulsivity	VDS	SNI	Increased impulsivity in right-sided SNI 1 month after surgery	[35]
Contextual memory	Fear conditioning/ extinction	SNI	Impaired contextual fear extension (starting at postsurgery day 10)	[98]
Memory (aversion)	IntelliCage	SNI	Prolonged maintenance of aversive memory in aged SNI mice 1 month after surgery	[135]
Working memory	Figure-8 maze/ T maze	SNI	Impaired learning performance in the initial time points (1–6 days) in a 3-week follow-up period	[136, 137]
Working memory	Figure-8 maze	Monoarthritis (CFA)	Impaired learning performance on the 4th and 5th postpain onset days in a 16-day follow-up period	[138]

Table 3. Continued

Trait	Paradigm	Model	Overall effect	Ref.
Spatial reference memory	Y-maze	SNL (L5)	Decreased % of time in the novel arm 2 weeks after surgery	[139]
Recognisance memory	NOR	EAE, monoarthritis (CFA)	EAE but not CFA animals failed to recognize the new object at day 7 post-disease induction	[140]
Social/working memory	Social recognition/ Y-maze	Monoarthritis (CFA)	Impairments in the Y-maze and social recognition paradigm 2–3 weeks after surgery	[108]
Memory	MWM/Y-maze/NOR	SCI	Impaired memory 2 months after surgery	[124]

Summary of the literature concerning the effects of chronic pain in cognitive performance in rodent models. Due to space constraints only positive results are shown. 8-arm = 8-arm radial maze; EAE = experimental autoimmune encephalomyelitis; CFA = complete Freund's adjuvant; K/C = kaolin/carrageenan; MWM = morris water maze; NOR = novel object recognition; PSNL = partial sciatic nerve ligation; SCI = spinal cord injury; SNI = spared nerve injury; SNL = spinal nerve ligation; WM = water maze.

ing and executive function, exceed by far 'emotional' behavior paradigms. Authors dealing with cognitive aspects in chronic pain have made use of this large pallet, making any systematization attempt very difficult. As in the emotional domain, several reviews have been published, including an extensive review by Finn and colleagues [52] that compiled the most relevant clinical and experimental works produced until 2011. In the clinical setting, attention, learning, memory, speed of information processing, psychomotor ability and executive function have been tested in several chronic pain conditions. Results are diverse and/or conflicting, which might be a result of either the different tools used to measure cognition, the specificities of the clinical conditions or both. Factors such as age, gender, pain duration and educational level (critical for the performance in many of the tasks) can account for much of the differences observed between studies. In the experimental studies using neuropathy models we face a similar scenario (for review, see Leite-Almeida et al. [53] and Low [54]), although the factors accounting for variability necessarily differ. These include model, experimental design (frequently a battery of tests is applied in tandem), paradigm construct and neuropathy side [35, 55]. An important (and often neglected) aspect of behavioral paradigms used to assess cognitive function is that they rely on a motor response; in a study from our group using the SNI model we clearly demonstrated an age-associated compromise in motor function [51]. The usage of travelled distance to the platform [51] or even dry mazes like the 8arm radial maze [see 'Nociceptive (Inflammatory) Pain'; [56]] in the SNI model showed in fact that this behavioral domain appears not to be affected at 1 month after pain onset. As in the 'emotional' domain, we also demonstrated that the age of the experimental subject or the injury side can be critical. For instance, while left-sided SNI has no effect on reversal learning of young subjects, it can be detrimental for mid-aged rats [51] or if in the right side [35]. In a recent study using the ASST, we demonstrated that the side of the peripheral lesion interfered with the interhemispheric activity balance in the orbitofrontal cortex [55], which might underlie side-specific deficits.

While in the 'emotional' domain an interrelation between pain and affect traits has been extensively demonstrated in humans and experimental models, such a relationship in the 'cognitive' domain has not been studied systematically. However, a recent study demonstrated that cognitive function can be a predictor of chronic pain [57], suggesting a similar bidirectional influence.

Nociceptive (Inflammatory) Pain

Models

According to the IASP (International Association for the Study of Pain), nociceptive pain corresponds to pain that arises from actual or threatened damage to nonneural tissue (IASP pain taxonomy), where the somatosensory nervous system functions normally. Nociceptive pain is considered somatic if pain originates from the joints, bones, muscles and other soft tissues, with arthritis comprising a group of heterogeneous diseases with different etiopathogenetic mechanisms ranging from progressive degeneration of cartilage (osteoarthritis) to an overactive immune system [rheumatoid arthritis (RA)] [58].

Experimental models have the great advantage of being able to mimic a large spectrum of clinically relevant features (table 1), but the biggest challenge remains the need to identify the critical events leading to the development and maintenance of human chronic pain.

RA models have contributed mainly to test therapeutic drugs that modulate events central to immunological regulation [59], but due to RA complexity most animal models only marginally reproduce the human disease. In osteoarthritic models, the lack of efficient treatment approaches is a major challenge. An important problem for animal models is the fact the human disease might take decades to develop and most researchers lack the facilities that would allow animal ageing [60]. Nonetheless, a great number of surgically induced models developed in species ranging from mice to horses have allowed many relevant questions to be answered and diagnosis techniques to be refined. The injection of irritants allowed investigation of pathophysiological mechanisms and treatment therapies [60]. Ultimately, however, there is an overall need to standardize preclinical studies by improving the uniformity of the factors controlled and reported, and their extrapolation to human diseases [60].

As shown in table 1, a myriad of animal models of arthritis were developed in the last decades. Many experimental models of inflammatory pain have been demonstrated to be robust, reproducible and easy to use, providing mechanistic insights and, more importantly, allowing prediction of the outcome of novel drug therapies serving as proof-of-principle experiments before clinical trials are approved.

Impact on Emotional Behavior

In humans, chronic pain is accompanied by altered emotions such as pain catastrophizing, anxiety, fear-avoidance behaviors and helplessness [61], and patients display higher levels of disability, healthcare usage, pain medication usage, attention to pain, depression, lower pain coping and longer rehabilitation times [61]. Nonetheless, only a small number of studies involving experimental models of arthritis have investigated the comorbidity between chronic pain and emotional impairments (table 2).

It is clear that experimental models of chronic inflammatory disorders develop comorbid emotional impairments such as anxiety and depression, similar to those reported in clinical settings. However, it is rather surprising that emotional impairments are studied almost exclusively in experimental models of osteoarthritis, and in particular in the complete Freund's adjuvant model, when many other animal models are available (table 2) and depression has a prevalence between 13 and 42% in RA patients [62] and around 50% in other rheumatic diseases such as systemic lupus erythematosus [63]. More importantly, many studies assessing anxiety-like behavior in animal models fail to evaluate, or to report, depressive-like phenotypes. This hiatus can partly be a consequence of the short time frames of disease settlement most commonly used, which contrast with the human condition that develops gradually over several years.

Overall the data in table 2 show an earlier onset of anxiety-like behaviors, some as soon as 1 day after induction, while depressive-like behavior is reported 2 weeks later. In fact, we have recently demonstrated [64] that by allowing the settlement of kaolin/carrageenan-induced monoarthritis for a period of 4 weeks, arthritic animals exhibited not only anxiety-like behavior as previously reported by other groups at earlier time frames such as 4 h [65], 5–6 h [66] and 1 week [67], but also the anhedonic-like and the learned helplessness-like components of depression.

Impact on Cognitive Performance

A review of the studies assessing cognition in models of arthritic pain is presented in table 3. As observed for neuropathic pain, a smaller number of studies evaluating cognitive performance in experimental models of arthritic pain have been published when compared to those evaluating emotional impairments. Once more, most studies were performed in an experimental model of osteoarthritis (complete Freund's adjuvant model) although 30% of RA patients display cognitive impairments [68] and this impairment is less significant in osteoarthritis.

The time allowed for osteoarthritis settlement also varied considerably between studies ranging from 5 h and 2 months, with animals consistently displaying attentional deficits and increased preference for risk soon after the onset of the model (from 5 h onwards) and memory impairments approximately 2 weeks later. Although cognitive decline in humans is a gradual process, studies show attentional deficits, reduced mental flexibility, executive function, visuospatial learning/memory and verbal learning/memory are present in 30–38% of RA patients [68, 69].

Dysfunctional Pain

Models

Dysfunctional pain refers to a group of clinical syndromes that are maladaptive and is caused by a malfunction of the somatosensory apparatus itself, as there is no identifiable noxious stimulus nor any detectable inflammation or damage to the nervous system (e.g. fibromyalgia, irritable bowel syndrome). In relation to dysfunctional pain models, those trying to reproduce different pathological alterations similar to fibromyalgia patients comprise 3 major types: animal models based on repeated muscle insult, biogenic amine depletion and stress [14].

The most studied model using repeated stimulation of the muscle is the acid saline-induced pain protocol [14, 70], where two acid saline injections separated by 2–5 days are given unilaterally to the gastrocnemius muscle. This results in a bilateral mechanical hyperalgesia that lasts for 4 weeks. As a major limitation, in these models hyperalgesia is not clearly associated with the development of comorbid anxiety, depression or fatigue. As a major advantage, pain can be reversed by different analgesic strategies that are commonly used to treat fibromyalgia patients, including drug protocols and physical (exercise) therapy. In another model, the hyperalgesic priming model, a series of insults to the muscle or skin is then followed by injection of prostaglandin E_2 to the same muscle, which results in hyperalgesia for at least 14 days. Major limitations in the present and in another model using a muscle-fatiguing task and a low-intensity muscle insult (fatigue-enhanced muscle pain model) are the absence of studies on comorbid mood disorders [14].

Impact on Emotional Behavior

The biogenic depletion model is the one matching common comorbid symptoms and pharmacological treatment profile for controlling chronic pain in patients with fibromyalgia (in male rats). The subcutaneous injection of reserpine for 3 days causes a sustained 1-week widespread muscle and cutaneous hyperalgesia (male and female rats) that is associated with a decreased amount of biogenic amines (dopamine, noradrenaline and serotonin) in the brain and spinal cord. Additionally, it is also associated with depressive-like comorbid symptoms as measured by the forced swimming test [14, 71].

Stress is the basis for the third group of animal models of fibromyalgia and the sound stress model induces increased anxiety [72].

Impact on Cognitive Performance

To the best of our knowledge, no data are available on comorbid pain and cognitive alterations in animal models of fibromyalgia [14]. However, since cognitive alterations are common in fibromyalgia patients, including decreased working memory, attention and executive functions [73], future studies should address these aspects in animal models.

Concluding Remarks

In the last 2 decades a myriad of behavioral paradigms have been introduced in the field of experimental pain research, particularly in chronic pain. These have encompassed tests of emotional and cognitive function, extending research beyond the boundaries of sensory aspects to comorbidities that frequently afflict chronic pain patients like depression, anxiety and impaired executive function. While there seems to be little doubt that chronic pain models manifest altered emotional and cognitive behaviors in a striking parallel with the human pathology, contradictory information abounds in the field. This stems from a large experimental variability including the use of different species/strains, pain models, time courses and lesion sites, but it might also embody biological phenomena that presently remain poorly understood.

References

1 Breivik H, Collett B, Ventafridda V, Cohen R, Gallacher D: Survey of chronic pain in Europe: prevalence, impact on daily life, and treatment. Eur J Pain 2006;10:287–333.

2 Schopflocher D, Taenzer P, Jovey R: The prevalence of chronic pain in Canada. Pain Res Manag 2011;16: 445–450.

3 Johannes CB, Le TK, Zhou X, Johnston JA, Dworkin RH: The prevalence of chronic pain in United States adults: results of an Internet-based survey. J Pain 2010;11:1230–1239.

4 Gaskin DJ, Richard P: The economic costs of pain in the United States. J Pain 2012;13:715–724.

5 McWilliams LA, Goodwin RD, Cox BJ: Depression and anxiety associated with three pain conditions: results from a nationally representative sample. Pain 2004;111:77–83.

6 Berryman C, Stanton TR, Jane Bowering K, Tabor A, McFarlane A, Lorimer Moseley G: Evidence for working memory deficits in chronic pain: a systematic review and meta-analysis. Pain 2013;154:1181–1196.

7 Liu MG, Chen J: Preclinical research on pain comorbidity with affective disorders and cognitive deficits: challenges and perspectives. Prog Neurobiol 2014; 116:13–32.

8 Bair MJ, Robinson RL, Katon W, Kroenke K: Depression and pain comorbidity: a literature review. Arch Intern Med 2003;163:2433–2445.

9 Mogil JS, Davis KD, Derbyshire SW: The necessity of animal models in pain research. Pain 2010;151:12–17.

10 Blackburn-Munro G: Pain-like behaviours in animals – how human are they? Trends Pharmacol Sci 2004;25:299–305.

11 Mogil JS: Animal models of pain: progress and challenges. Nat Rev Neurosci 2009;10:283–294.

12 Le Bars D, Gozariu M, Cadden SW: Animal models of nociception. Pharmacol Rev 2001;53:597–652.

13 Barrot M: Tests and models of nociception and pain in rodents. Neuroscience 2012;211:39–50.

14 DeSantana JM, da Cruz KM, Sluka KA: Animal models of fibromyalgia. Arthritis Res Ther 2013;15: 222.

15 Whiteside GT, Adedoyin A, Leventhal L: Predictive validity of animal pain models? A comparison of the pharmacokinetic-pharmacodynamic relationship for pain drugs in rats and humans. Neuropharmacology 2008;54:767–775.

16 Kontinen VK, Kauppila T, Paananen S, Pertovaara A, Kalso E: Behavioural measures of depression and anxiety in rats with spinal nerve ligation-induced neuropathy. Pain 1999;80:341–346.

17 Damasio A, Carvalho GB: The nature of feelings: evolutionary and neurobiological origins. Nat Rev Neurosci 2013;14:143–152.

18 Sousa N, Almeida OF, Wotjak CT: A hitchhiker's guide to behavioral analysis in laboratory rodents. Genes Brain Behav 2006;5(suppl 2):5–24.

19 Shepherd JK, Grewal SS, Fletcher A, Bill DJ, Dourish CT: Behavioural and pharmacological characterisation of the elevated 'zero-maze' as an animal model of anxiety. Psychopharmacology (Berl) 1994;116:56–64.

20 Hall CS: Emotional behavior in the rat. I. Defecation and urination as measures of individual differences in emotionality. J Comp Psychol 1934;18:385–403.

21 Crawley JN: Neuropharmacologic specificity of a simple animal model for the behavioral actions of benzodiazepines. Pharmacol Biochem Behav 1981; 15:695–699.

22 LaBuda CJ, Fuchs PN: A behavioral test paradigm to measure the aversive quality of inflammatory and neuropathic pain in rats. Exp Neurol 2000;163:490–494.
23 Porsolt RD, Le Pichon M, Jalfre M: Depression: a new animal model sensitive to antidepressant treatments. Nature 1977;266:730–732.
24 Steru L, Chermat R, Thierry B, Simon P: The tail suspension test: a new method for screening antidepressants in mice. Psychopharmacology (Berl) 1985;85:367–370.
25 Mateus-Pinheiro A, Patricio P, Alves ND, Machado-Santos AR, Morais M, Bessa JM, et al: The Sweet Drive Test: refining phenotypic characterization of anhedonic behavior in rodents. Front Behav Neurosci 2014;8:74.
26 Deacon RM: Burrowing in rodents: a sensitive method for detecting behavioral dysfunction. Nat Protoc 2006;1:118–121.
27 Rutten K, Schiene K, Robens A, Leipelt A, Pasqualon T, Read SJ, et al: Burrowing as a non-reflex behavioural readout for analgesic action in a rat model of sub-chronic knee joint inflammation. Eur J Pain 2014;18:204–212.
28 Andrews N, Legg E, Lisak D, Issop Y, Richardson D, Harper S, et al: Spontaneous burrowing behaviour in the rat is reduced by peripheral nerve injury or inflammation associated pain. Eur J Pain 2012;16:485–495.
29 Morris R: Developments of a water-maze procedure for studying spatial learning in the rat. J Neurosci Methods 1984;11:47–60.
30 Birrell JM, Brown VJ: Medial frontal cortex mediates perceptual attentional set shifting in the rat. J Neurosci 2000;20:4320–4324.
31 Leonard JA: 5-Choice Serial Reaction Apparatus (Applied Psychology Unit Report 326/59). Cambridge, Medical Research Council 1959.
32 Bari A, Dalley JW, Robbins TW: The application of the 5-choice serial reaction time task for the assessment of visual attentional processes and impulse control in rats. Nat Protoc 2008;3:759–767.
33 Pais-Vieira M, Lima D, Galhardo V: Sustained attention deficits in rats with chronic inflammatory pain. Neurosci Lett 2009;463:98–102.
34 Leite-Almeida H, Melo A, Pego JM, Bernardo S, Milhazes N, Borges F, et al: Variable delay-to-signal: a fast paradigm for assessment of aspects of impulsivity in rats. Front Behav Neurosci 2013;7:154.
35 Leite-Almeida H, Cerqueira JJ, Wei H, Ribeiro-Costa N, Anjos-Martins H, Sousa N, et al: Differential effects of left/right neuropathy on rats' anxiety and cognitive behavior. Pain 2012;153:2218–2225.
36 Costigan M, Scholz J, Woolf CJ: Neuropathic pain: a maladaptive response of the nervous system to damage. Annu Rev Neurosci 2009;32:1–32.
37 Handwerker HO, Arendt-Neilsen L: Pain Models: Translational Relevance and Applications, ed 1. Seattle, IASP Press, 2013.
38 Jensen TS, Baron R, Haanpaa M, Kalso E, Loeser JD, Rice AS, et al: A new definition of neuropathic pain. Pain 2011;152:2204–2205.
39 Decosterd I, Berta T: Animal models and neuropathic pain; in Basbaum AI, Bushnell MC (eds): Science of Pain. Oxford, Academic Press, 2009, pp 857–864.
40 Yalcin I, Barthas F, Barrot M: Emotional consequences of neuropathic pain: insight from preclinical studies. Neurosci Biobehav Rev 2014;47C:154–164.
41 Hasnie FS, Breuer J, Parker S, Wallace V, Blackbeard J, Lever I, et al: Further characterization of a rat model of varicella zoster virus-associated pain: relationship between mechanical hypersensitivity and anxiety-related behavior, and the influence of analgesic drugs. Neuroscience 2007;144:1495–1508.
42 Roeska K, Doods H, Arndt K, Treede RD, Ceci A: Anxiety-like behaviour in rats with mononeuropathy is reduced by the analgesic drugs morphine and gabapentin. Pain 2008;139:349–357.
43 Roeska K, Ceci A, Treede RD, Doods H: Effect of high trait anxiety on mechanical hypersensitivity in male rats. Neurosci Lett 2009;464:160–164.
44 Wiech K, Tracey I: The influence of negative emotions on pain: behavioral effects and neural mechanisms. Neuroimage 2009;47:987–994.
45 Yalcin I, Bohren Y, Waltisperger E, Sage-Ciocca D, Yin JC, Freund-Mercier MJ, et al: A time-dependent history of mood disorders in a murine model of neuropathic pain. Biol Psychiatry 2011;70:946–953.
46 Suzuki T, Amata M, Sakaue G, Nishimura S, Inoue T, Shibata M, et al: Experimental neuropathy in mice is associated with delayed behavioral changes related to anxiety and depression. Anesth Analg 2007;104:1570–1577.
47 Seminowicz DA, Laferriere AL, Millecamps M, Yu JS, Coderre TJ, Bushnell MC: MRI structural brain changes associated with sensory and emotional function in a rat model of long-term neuropathic pain. Neuroimage 2009;47:1007–1014.
48 Geha PY, Baliki MN, Harden RN, Bauer WR, Parrish TB, Apkarian AV: The brain in chronic CRPS pain: abnormal gray-white matter interactions in emotional and autonomic regions. Neuron 2008;60:570–581.
49 Sullivan RM, Gratton A: Prefrontal cortical regulation of hypothalamic-pituitary-adrenal function in the rat and implications for psychopathology: side matters. Psychoneuroendocrinology 2002;27:99–114.
50 Gagliese L, Schiff BB, Taylor A: Differential consequences of left- and right-sided chronic pain. Clin J Pain 1995;11:201–207.

51 Leite-Almeida H, Almeida-Torres L, Mesquita AR, Pertovaara A, Sousa N, Cerqueira JJ, et al: The impact of age on emotional and cognitive behaviours triggered by experimental neuropathy in rats. Pain 2009;144:57–65.
52 Moriarty O, McGuire BE, Finn DP: The effect of pain on cognitive function: a review of clinical and preclinical research. Prog Neurobiol 2011;93:385–404.
53 Leite-Almeida H, Pertovaara A, Almeida A: Emotional and cognitive impairments in experimental neuropathies. 4th International Congress on Neuropathic Pain, Bologna, 2013, pp 39–42.
54 Low LA: The impact of pain upon cognition: what have rodent studies told us? Pain 2013;154:2603–2605.
55 Leite-Almeida H, Guimaraes MR, Cerqueira JJ, Ribeiro-Costa N, Anjos-Martins H, Sousa N, et al: Asymmetric c-fos expression in the ventral orbital cortex is associated with impaired reversal learning in a right-sided neuropathy. Mol Pain 2014;10:41.
56 Ren WJ, Liu Y, Zhou LJ, Li W, Zhong Y, Pang RP, et al: Peripheral nerve injury leads to working memory deficits and dysfunction of the hippocampus by upregulation of TNF-alpha in rodents. Neuropsychopharmacology 2011;36:979–992.
57 Attal N, Masselin-Dubois A, Martinez V, Jayr C, Albi A, Fermanian J, et al: Does cognitive functioning predict chronic pain? Results from a prospective surgical cohort. Brain 2014;137:904–917.
58 Di Paola R, Cuzzocrea S: Predictivity and sensitivity of animal models of arthritis. Autoimmun Rev 2008; 8:73–75.
59 Dinser R: Animal models for arthritis. Best Pract Res Clin Rheumatol 2008;22:253–267.
60 Little CB, Zaki S: What constitutes an 'animal model of osteoarthritis' – the need for consensus? Osteoarthr Cartilage 2012;20:261–267.
61 Keefe FJ, Rumble ME, Scipio CD, Giordano LA, Perri LM: Psychological aspects of persistent pain: current state of the science. J Pain 2004;5:195–211.
62 Isik A, Koca SS, Ozturk A, Mermi O: Anxiety and depression in patients with rheumatoid arthritis. Clin Rheumatol 2007;26:872–878.
63 Kozora E, Ellison MC, West S: Depression, fatigue, and pain in systemic lupus erythematosus (SLE): relationship to the American College of Rheumatology SLE neuropsychological battery. Arthritis Rheum 2006;55:628–635.
64 Amorim D, David-Pereira A, Pertovaara A, Almeida A, Pinto-Ribeiro F: Amitriptyline reverses hyperalgesia and improves associated mood-like disorders in a model of experimental monoarthritis. Behav Brain Res 2014;265:12–21.
65 Lu Y, Westlund KN: Gabapentin attenuates nociceptive behaviors in an acute arthritis model in rats. J Pharmacol Exp Ther 1999;290:214–219.
66 Ji G, Fu Y, Ruppert KA, Neugebauer V: Pain-related anxiety-like behavior requires CRF1 receptors in the amygdala. Mol Pain 2007;3:13.
67 Kim KS, Kim MH, Yeom M, Choi HM, Yang HI, Yoo MC, et al: Arthritic disease is more severe in older rats in a kaolin/carrageenan-induced arthritis model. Rheumatol Int 2012;32:3875–3879.
68 Appenzeller S, Bertolo MB, Costallat LT: Cognitive impairment in rheumatoid arthritis. Methods Find Exp Clin Pharmacol 2004;26:339–343.
69 Bartolini M, Candela M, Brugni M, Catena L, Mari F, Pomponio G, et al: Are behaviour and motor performances of rheumatoid arthritis patients influenced by subclinical cognitive impairments? A clinical and neuroimaging study. Clin Exp Rheumatol 2002;20:491–497.
70 Sluka KA, Kalra A, Moore SA: Unilateral intramuscular injections of acidic saline produce a bilateral, long-lasting hyperalgesia. Muscle Nerve 2001;24:37–46.
71 Nagakura Y, Oe T, Aoki T, Matsuoka N: Biogenic amine depletion causes chronic muscular pain and tactile allodynia accompanied by depression: a putative animal model of fibromyalgia. Pain 2009;146: 26–33.
72 Green PG, Alvarez P, Gear RW, Mendoza D, Levine JD: Further validation of a model of fibromyalgia syndrome in the rat. J Pain 2011;12:811–818.
73 Bertolucci PH, de Oliveira FF: Cognitive impairment in fibromyalgia. Curr Pain Headache Rep 2013;17: 344.
74 Klusakova I, Dubovy P: Experimental models of peripheral neuropathic pain based on traumatic nerve injuries – an anatomical perspective. Ann Anat 2009; 191:248–259.
75 Bennett GJ: Pathophysiology and animal models of cancer-related painful peripheral neuropathy. Oncologist 2010;15(suppl 2):9–12.
76 Willison HJ: Biomarkers in experimental models of antibody-mediated neuropathies. J Peripher Nerv Syst 2011;16(suppl 1):60–62.
77 Calcutt NA: Experimental models of painful diabetic neuropathy. J Neurol Sci 2004;220:137–139.
78 Meyer Zu Horste G, Nave KA: Animal models of inherited neuropathies. Curr Opin Neurol 2006;19: 464–473.
79 Authier N, Balayssac D, Marchand F, Ling B, Zangarelli A, Descoeur J, et al: Animal models of chemotherapy-evoked painful peripheral neuropathies. Neurotherapeutics 2009;6:620–629.
80 Nakae A, Nakai K, Yano K, Hosokawa K, Shibata M, Mashimo T: The animal model of spinal cord injury as an experimental pain model. J Biomed Biotechnol 2011;2011:939023.
81 Teeple E, Jay GD, Elsaid KA, Fleming BC: Animal models of osteoarthritis: challenges of model selection and analysis. AAPS J 2013;15:438–446.

82 Naik SR, Wala SM: Arthritis, a complex connective and synovial joint destructive autoimmune disease: animal models of arthritis with varied etiopathology and their significance. J Postgrad Med 2014;60:309–317.
83 Longo UG, Loppini M, Fumo C, Rizzello G, Khan WS, Maffulli N, et al: Osteoarthritis: new insights in animal models. Open Orthop J 2012;6:558–563.
84 Kannan K, Ortmann RA, Kimpel D: Animal models of rheumatoid arthritis and their relevance to human disease. Pathophysiology 2005;12:167–181.
85 Fernandez-Guasti A, Reyes R, Martinez-Mota L, Lopez-Munoz FJ: Influence of inflammatory nociception on the anxiolytic-like effect of diazepam and buspirone in rats. Psychopharmacology (Berl) 2005;180:399–407.
86 Narita M, Kaneko C, Miyoshi K, Nagumo Y, Kuzumaki N, Nakajima M, et al: Chronic pain induces anxiety with concomitant changes in opioidergic function in the amygdala. Neuropsychopharmacology 2006;31:739–750.
87 Narita M, Kuzumaki N, Narita M, Kaneko C, Hareyama N, Miyatake M, et al: Chronic pain-induced emotional dysfunction is associated with astrogliosis due to cortical delta-opioid receptor dysfunction. J Neurochem 2006;97:1369–1378.
88 Wallace VC, Blackbeard J, Pheby T, Segerdahl AR, Davies M, Hasnie F, et al: Pharmacological, behavioural and mechanistic analysis of HIV-1 gp120 induced painful neuropathy. Pain 2007;133:47–63.
89 Wallace VC, Segerdahl AR, Blackbeard J, Pheby T, Rice AS: Anxiety-like behaviour is attenuated by gabapentin, morphine and diazepam in a rodent model of HIV anti-retroviral-associated neuropathic pain. Neurosci Lett 2008;448:153–156.
90 Wallace VC, Blackbeard J, Segerdahl AR, Hasnie F, Pheby T, McMahon SB, et al: Characterization of rodent models of HIV-gp120 and anti-retroviral-associated neuropathic pain. Brain 2007;130:2688–2702.
91 Benbouzid M, Pallage V, Rajalu M, Waltisperger E, Doridot S, Poisbeau P, et al: Sciatic nerve cuffing in mice: a model of sustained neuropathic pain. Eur J Pain 2008;12:591–599.
92 Matsuzawa-Yanagida K, Narita M, Nakajima M, Kuzumaki N, Niikura K, Nozaki H, et al: Usefulness of antidepressants for improving the neuropathic pain-like state and pain-induced anxiety through actions at different brain sites. Neuropsychopharmacology 2007;33:1952–1965.
93 Benatti C, Alboni S, Capone G, Corsini D, Caggia F, Brunello N, et al: Early neonatal inflammation affects adult pain reactivity and anxiety related traits in mice: genetic background counts. Int J Dev Neurosci 2009;27:661–668.
94 Griffin TM, Fermor B, Huebner JL, Kraus VB, Rodriguiz RM, Wetsel WC, et al: Diet-induced obesity differentially regulates behavioral, biomechanical, and molecular risk factors for osteoarthritis in mice. Arthritis Res Ther 2010;12:R130.
95 Urban R, Scherrer G, Goulding EH, Tecott LH, Basbaum AI: Behavioral indices of ongoing pain are largely unchanged in male mice with tissue or nerve injury-induced mechanical hypersensitivity. Pain 2011;152:990–1000.
96 Gregoire S, Michaud V, Chapuy E, Eschalier A, Ardid D: Study of emotional and cognitive impairments in mononeuropathic rats: effect of duloxetine and gabapentin. Pain 2012;153:1657–1663.
97 Kim H, Chen L, Lim G, Sung B, Wang S, McCabe MF, et al: Brain indoleamine 2,3-dioxygenase contributes to the comorbidity of pain and depression. J Clin Invest 2012;122:2940–2954.
98 Mutso AA, Radzicki D, Baliki MN, Huang L, Banisadr G, Centeno MV, et al: Abnormalities in hippocampal functioning with persistent pain. J Neurosci 2012;32:5747–5756.
99 Parent AJ, Beaudet N, Beaudry H, Bergeron J, Berube P, Drolet G, et al: Increased anxiety-like behaviors in rats experiencing chronic inflammatory pain. Behav Brain Res 2012;229:160–167.
100 Alba-Delgado C, Llorca-Torralba M, Horrillo I, Ortega JE, Mico JA, Sanchez-Blazquez P, et al: Chronic pain leads to concomitant noradrenergic impairment and mood disorders. Biol Psychiatry 2013;73:54–62.
101 do Nascimento GC, Leite-Panissi CR: Time-dependent analysis of nociception and anxiety-like behavior in rats submitted to persistent inflammation of the temporomandibular joint. Physiol Behav 2014;125:1–7.
102 Chen J, Song Y, Yang J, Zhang Y, Zhao P, Zhu XJ, et al: The contribution of TNF-alpha in the amygdala to anxiety in mice with persistent inflammatory pain. Neurosci Lett 2013;541:275–280.
103 Avila-Martin G, Galan-Arriero I, Ferrer-Donato A, Busquets X, Gomez-Soriano J, Escriba PV, et al: Oral 2-hydroxyoleic acid inhibits reflex hypersensitivity and open-field-induced anxiety after spared nerve injury. Eur J Pain 2015;19:111–122.
104 Borges G, Neto F, Mico JA, Berrocoso E: Reversal of monoarthritis-induced affective disorders by diclofenac in rats. Anesthesiology 2014;120:1476–1490.
105 Caspani O, Reitz MC, Ceci A, Kremer A, Treede RD: Tramadol reduces anxiety-related and depression-associated behaviors presumably induced by pain in the chronic constriction injury model of neuropathic pain in rats. Pharmacol Biochem Behav 2014;124:290–296.

106 Dimitrov EL, Tsuda MC, Cameron HA, Usdin TB: Anxiety- and depression-like behavior and impaired neurogenesis evoked by peripheral neuropathy persist following resolution of prolonged tactile hypersensitivity. J Neurosci 2014;34:12304–12312.
107 Galan-Arriero I, Avila-Martin G, Ferrer-Donato A, Gomez-Soriano J, Bravo-Esteban E, Taylor J: Oral administration of the p38alpha MAPK inhibitor, UR13870, inhibits affective pain behavior after spinal cord injury. Pain 2014;155:2188–2198.
108 Gregoire S, Wattiez AS, Etienne M, Marchand F, Ardid D: Monoarthritis-induced emotional and cognitive impairments in rats are sensitive to low systemic doses or intra-amygdala injections of morphine. Eur J Pharmacol 2014;735:1–9.
109 Hache G, Guiard BP, Nguyen TH, Quesseveur G, Gardier AM, Peters D, et al: Antinociceptive activity of the new triple reuptake inhibitor NS18283 in a mouse model of chemotherapy-induced neuropathic pain. Eur J Pain 2015;19:322–333.
110 Han M, Xiao X, Yang Y, Huang RY, Cao H, Zhao ZQ, et al: SIP30 is required for neuropathic pain-evoked aversion in rats. J Neurosci 2014;34:346–355.
111 Li Q, Yue N, Liu SB, Wang ZF, Mi WL, Jiang JW, et al: Effects of chronic electroacupuncture on depression- and anxiety-like behaviors in rats with chronic neuropathic pain. Evid Based Complement Alternat Med 2014;2014:158987.
112 Sawada A, Niiyama Y, Ataka K, Nagaishi K, Yamakage M, Fujimiya M: Suppression of bone marrow-derived microglia in the amygdala improves anxiety-like behavior induced by chronic partial sciatic nerve ligation in mice. Pain 2014;155:1762–1772.
113 Gonçalves L, Silva R, Pinto-Ribeiro F, Pego JM, Bessa JM, Pertovaara A, et al: Neuropathic pain is associated with depressive behaviour and induces neuroplasticity in the amygdala of the rat. Exp Neurol 2008;213:48–56.
114 Hu B, Doods H, Treede RD, Ceci A: Depression-like behaviour in rats with mononeuropathy is reduced by the CB2-selective agonist GW405833. Pain 2009;143:206–212.
115 Norman GJ, Karelina K, Zhang N, Walton JC, Morris JS, Devries AC: Stress and IL-1beta contribute to the development of depressive-like behavior following peripheral nerve injury. Mol Psychiatry 2009;15:404–414.
116 Hu Y, Yang J, Wang Y, Li W: Amitriptyline rather than lornoxicam ameliorates neuropathic pain-induced deficits in abilities of spatial learning and memory. Eur J Anaesthesiol 2010;27:162–168.
117 Fukuhara K, Ishikawa K, Yasuda S, Kishishita Y, Kim HK, Kakeda T, et al: Intracerebroventricular 4-methylcatechol (4-MC) ameliorates chronic pain associated with depression-like behavior via induction of brain-derived neurotrophic factor (BDNF). Cell Mol Neurobiol 2012;32:971–977.
118 Bura AS, Guegan T, Zamanillo D, Vela JM, Maldonado R: Operant self-administration of a sigma ligand improves nociceptive and emotional manifestations of neuropathic pain. Eur J Pain 2013;17:832–843.
119 Goffer Y, Xu D, Eberle SE, D'Amour J, Lee M, Tukey D, et al: Calcium-permeable AMPA receptors in the nucleus accumbens regulate depression-like behaviors in the chronic neuropathic pain state. J Neurosci 2013;33:19034–19044.
120 Maciel IS, Silva RB, Morrone FB, Calixto JB, Campos MM: Synergistic effects of celecoxib and bupropion in a model of chronic inflammation-related depression in mice. PLoS One 2013;8:e77227.
121 Stratinaki M, Varidaki A, Mitsi V, Ghose S, Magida J, Dias C, et al: Regulator of G protein signaling is a crucial modulator of antidepressant drug action in depression and neuropathic pain models. Proc Natl Acad Sci U S A 2013;110:8254–8259.
122 Dellarole A, Morton P, Brambilla R, Walters W, Summers S, Bernardes D, et al: Neuropathic pain-induced depressive-like behavior and hippocampal neurogenesis and plasticity are dependent on TNFR1 signaling. Brain Behav Immun 2014;41:65–81.
123 Gai BM, Bortolatto CF, Bruning CA, Zborowski VA, Stein AL, Zeni G, et al: Depression-related behavior and mechanical allodynia are blocked by 3-(4-fluorophenylselenyl)-2,5-diphenylselenophene in a mouse model of neuropathic pain induced by partial sciatic nerve ligation. Neuropharmacology 2014;79:580–589.
124 Wu J, Zhao Z, Sabirzhanov B, Stoica BA, Kumar A, Luo T, et al: Spinal cord injury causes brain inflammation associated with cognitive and affective changes: role of cell cycle pathways. J Neurosci 2014;34:10989–11006.
125 Zhao X, Wang C, Zhang JF, Liu L, Liu AM, Ma Q, et al: Chronic curcumin treatment normalizes depression-like behaviors in mice with mononeuropathy: involvement of supraspinal serotonergic system and GABAA receptor. Psychopharmacology (Berl) 2014;231:2171–2187.
126 Zhao X, Yu C, Wang C, Zhang JF, Zhou WH, Cui WG, et al: Chronic resveratrol treatment exerts antihyperalgesic effect and corrects co-morbid depressive like behaviors in mice with mononeuropathy: involvement of serotonergic system. Neuropharmacology 2014;85:131–141.

127 Cain CK, Francis JM, Plone MA, Emerich DF, Lindner MD: Pain-related disability and effects of chronic morphine in the adjuvant-induced arthritis model of chronic pain. Physiol Behav 1997;62: 199–205.
128 Lindner MD, Plone MA, Francis JM, Cain CK: Chronic morphine reduces pain-related disability in a rodent model of chronic, inflammatory pain. Exp Clin Psychopharmacol 1999;7:187–197.
129 Millecamps M, Etienne M, Jourdan D, Eschalier A, Ardid D: Decrease in non-selective, non-sustained attention induced by a chronic visceral inflammatory state as a new pain evaluation in rats. Pain 2004;109:214–224.
130 Zhao MG, Ko SW, Wu LJ, Toyoda H, Xu H, Quan J, et al: Enhanced presynaptic neurotransmitter release in the anterior cingulate cortex of mice with chronic pain. J Neurosci 2006;26:8923–8930.
131 Hummel M, Lu PM, Cummons TA, Whiteside GT: The persistence of a long-term negative affective state following the induction of either acute or chronic pain. Pain 2008;140:436–445.
132 Pais-Vieira M, Mendes-Pinto MM, Lima D, Galhardo V: Cognitive impairment of prefrontal-dependent decision-making in rats after the onset of chronic pain. Neuroscience 2009;161:671–679.
133 Ji G, Sun H, Fu Y, Li Z, Pais-Vieira M, Galhardo V, et al: Cognitive impairment in pain through amygdala-driven prefrontal cortical deactivation. J Neurosci 2010;30:5451–5464.
134 Kodama D, Ono H, Tanabe M: Increased hippocampal glycine uptake and cognitive dysfunction after peripheral nerve injury. Pain 2011;152:809–817.
135 Albuquerque B, Haussler A, Vannoni E, Wolfer DP, Tegeder I: Learning and memory with neuropathic pain: impact of old age and progranulin deficiency. Front Behav Neurosci 2013;7:174.
136 Cardoso-Cruz H, Lima D, Galhardo V: Impaired spatial memory performance in a rat model of neuropathic pain is associated with reduced hippocampus-prefrontal cortex connectivity. J Neurosci 2013;33:2465–2480.
137 Cardoso-Cruz H, Dourado M, Monteiro C, Matos MR, Galhardo V: Activation of dopaminergic D2/D3 receptors modulates dorsoventral connectivity in the hippocampus and reverses the impairment of working memory after nerve injury. J Neurosci 2014;34:5861–5873.
138 Cardoso-Cruz H, Sousa M, Vieira JB, Lima D, Galhardo V: Prefrontal cortex and mediodorsal thalamus reduced connectivity is associated with spatial working memory impairment in rats with inflammatory pain. Pain 2013;154:2397–2406.
139 Morel V, Etienne M, Wattiez AS, Dupuis A, Privat AM, Chalus M, et al: Memantine, a promising drug for the prevention of neuropathic pain in rat. Eur J Pharmacol 2013;721:382–390.
140 Olechowski CJ, Tenorio G, Sauve Y, Kerr BJ: Changes in nociceptive sensitivity and object recognition in experimental autoimmune encephalomyelitis (EAE). Exp Neurol 2013;241:113–121.

Prof. Armando Almeida
Life and Health Sciences Research Institute, School of Health Sciences
Campus de Gualtar, University of Minho
PO–4710-057 Braga (Portugal)
E-Mail aalmeida@ecsaude.uminho.pt

Finn DP, Leonard BE (eds): Pain in Psychiatric Disorders. Mod Trends Pharmacopsychiatry.
Basel, Karger, 2015, vol 30, pp 22–35 (DOI: 10.1159/000435930)

Pain, Depression and Inflammation: Are Interconnected Causative Factors Involved?

Brian E. Leonard

Department of Pharmacology, National University of Ireland, Galway, Ireland

Abstract

Co-morbid depression and chronic pain are highly prevalent. The purpose of this review is to examine the role of chronic inflammation as a common mediator of these co-morbidities. Dysfunctional bidirectional pathways between the brain and the immune, endocrine and neurotransmitter systems have been extensively described and implicated in pain and psychiatric disorders. This short review therefore accesses the evidence in favour of the psychoneuroendocrine hypothesis of psychiatric disorders under three main headings: (1) by illustrating how different types of stress play a crucial role in initiating chronic inflammation in major depression, (2) by accessing the evidence that pain is frequently an important component of, and an initiator of, depression, and (3) considering the evidence that chronic inflammation provides an important link between chronic pain and depression, and the possible cellular mechanisms involved in this process. By understanding the critical role that chronic inflammation plays in pain and depression, novel approaches to the development of drugs may emerge that offer improvements in treatment.

The causes, expression and treatment of pain have preoccupied philosophers, theologians and clinicians since antiquity. Today, there is an ever-growing interest in the management and treatment of pain, linked to an exponential rise in basic and clinical research into the mechanisms whereby pain is initiated, propagated and terminated.

Historically, the early Christian theologian St. Augustine (354–430) recognised that 'the greatest evil is physical pain' while, a few centuries later, the great English writer John Milton (1608–74), in his *Paradise Lost* stated 'But pain is miserie, the worst of evils, and excessive, overturns all patience'.

It is interesting to note that the emphasis of these writers is on physical pain, as distinct from a more modern definition of pain, defined in the *Oxford English Dictionary* as 'suffering or distress of body or mind'. This definition is particularly pertinent

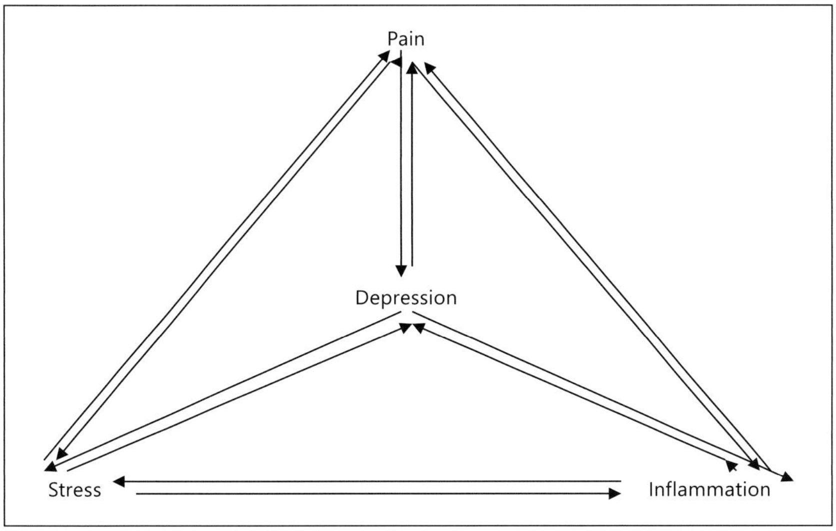

Fig. 1. Summary of the functional triad linking pain, stress and inflammation with chronic depression.

to the current chapter as co-morbid depression and pain occur with a high prevalence [1]. However, there has been relatively little research conducted into the relationship between inflammation, pain and depression, which will form the topic of this chapter (fig. 1).

However, in addition to the conventional treatment of chronic pain with non-steroidal anti-inflammatory drugs or opioids, antidepressants are now widely used even in first-line treatment. The rationale for the use of most classes of antidepressants is related to their effects on the monoamine neurotransmitters, which are located in the primary pain pathways in the brain [2]. Furthermore, it is now evident that there is an overlap between the neurochemical markers of pain development and depression [3]. This inter-relationship between pain, depression and antidepressants is illustrated in figure 2.

Stress, Inflammation and Depression

Activation of the immune system leading to inflammation is an adaptive response developed in the course of evolution to protect the organism from the harmful effects of pathogens, allergens and oncogenes. Hitherto, it was thought that the brain was an immune-privileged organ which played only a small role in the immune response. However, it is now accepted that there is a bidirectional relationship between the brain and the periphery by which neurons and immune cells are in close proximity. This is illustrated by the parenteral administration of a bacterial endotoxin to an experimental

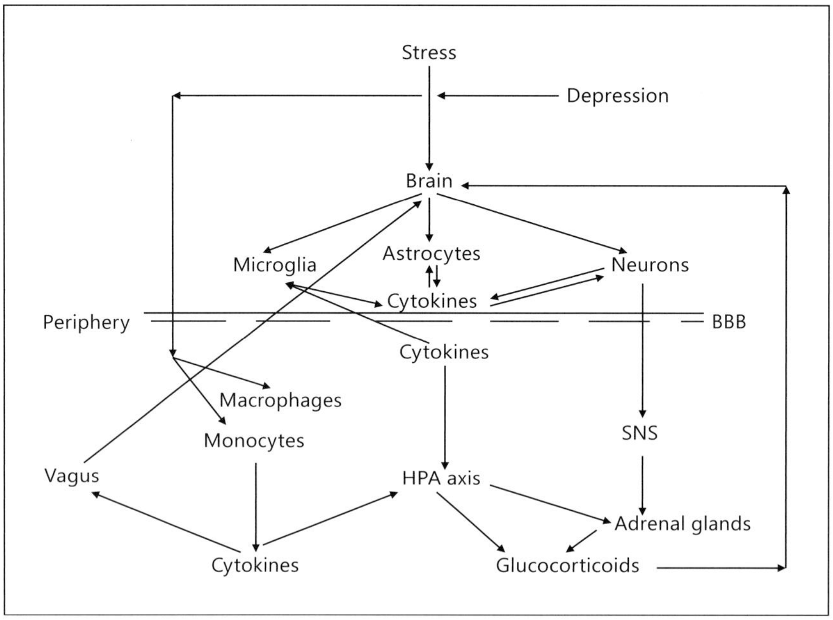

Fig. 2. Bidirectional connections between stress, brain and inflammatory cytokines. Afferent pathways: central and peripheral macrophages/microglia are activated by stress, pain and depression. Efferent pathways: glucocorticoids and proinflammatory cytokines activate the immune system in the brain via the vagus nerve and the endothelial cells which line the blood-brain barrier and signal the activation of the microglia. BBB = Blood-brain barrier; SNS = sympathetic nervous system.

animal or a human volunteer which results in the activation of the innate immune system, thereby triggering a cascade of immune changes that affect both the brain and peripheral organs [4]. While it is readily apparent that bacterial endotoxins initiate an acute inflammatory response, which has a protective function, it is only in more recent times that the maladaptive consequences of chronic inflammation, as initiated by environmental stressors for example, have received due attention.

The invasion of pathogens results in a number of symptoms that characterise sickness behaviour. These include fatigue, anorexia, reduction in motivation and apathy, loss of libido, and sleep disturbance. Many of these symptoms are also found in chronic fatigue syndrome in which there is little evidence that pathogens are involved. This suggests that sickness behaviour, in which the release of proinflammatory cytokines plays a major role, has an important function in encouraging rest and recuperation. An increased sensitivity to pain also enhances this behaviour.

The occurrence of the symptoms of sickness behaviour, and a reduction in the pain threshold, also occur early in the onset of major depression. This has led Maes et al. [5], and others [6, 7], to conclude that shared immunoinflammatory pathways underpin the pathophysiology of major depression, thereby explaining why the symptoms overlap. Inflammation, whether due to an infection, environmental stressor or other causes, can therefore be postulated to have a potentially beneficial acute protective role but a detrimental effect on physical and mental health if it becomes chronic. Maes et al. [8] provided crucial evidence linking proinflammatory cytokines to major depression by showing that the therapeutic use of such cytokines in non-depressed individuals with systemic diseases such as hepatitis C or malignant melanoma resulted in the symptoms of depression in a high proportion of the patients. This was further substantiated by Capuron et al. [9], who also demonstrated that chronic antidepressant treatments were more effective in attenuating the depressive symptoms which occurred later in the therapy than those that occurred during the sickness behaviour phase when the antidepressants had little beneficial effect.

Experimental and clinical studies have demonstrated that chronic stress activates the innate immune system and thereby triggers a cascade of inflammatory changes that impact on the brain. In the periphery, stress-activated macrophages and monocytes release a number of proinflammatory cytokines [e.g. interleukin (IL)-1 and -6, tumour necrosis factor (TNF)-α and interferon (IFN)-γ]. In the brain, similar cytokines are produced by stress-activated microglia and astrocytes, but an additional and important source of proinflammatory cytokines in the brain results from the impact of the peripheral cytokines on the endothelial cells of the blood-brain barrier – these signal changes to the microglia [10, 11]. The peripheral derived cytokines can also enhance the inflammatory changes to the brain by activating the peripheral terminal of the vagus nerve [12]. There is also experimental evidence that the proinflammatory cytokine IL-1α is actively transported across the blood-brain barrier [13].

The 'cross-talk' between the immune changes in the brain and the peripheral organs involves the central sympathetic and parasympathetic systems, which directly influence peripheral immune cells by activating the adrenergic and cholinergic receptors located on their outer membranes. In addition, the inflammatory changes also activate the hypothalamic-pituitary-adrenal (HPA) axis and thereby the secretion of the predominantly anti-inflammatory glucocorticoids from the adrenal cortex. Glucocorticoid receptors are widely distributed on immune cells and neurons in the brain, thereby permitting important 'feedback' between the endocrine and immune systems and the brain. Thus, the neuronal and humoral processes are involved in the coordination of the physiological, immunological and behavioural responses [14]. Figure 3 summarises the bidirectional pathways between the immune, endocrine and neurotransmitter systems.

Just as sickness behaviour is an important adaptive factor to recovery from an infection, the associated activation of the HPA axis, a component of the 'fight or flight' response of Cannon, has an important protective role. However, when the activation of the HPA axis is prolonged and unconnected with an appropriate stressful stimulus, the consequences are detrimental to both the brain and the peripheral organs. This situation occurs in major depression as approximately 50% of depressed patients exhibit hyperactivity of the HPA axis, a situation which increases to approximately 80% in those who are severely depressed [15].

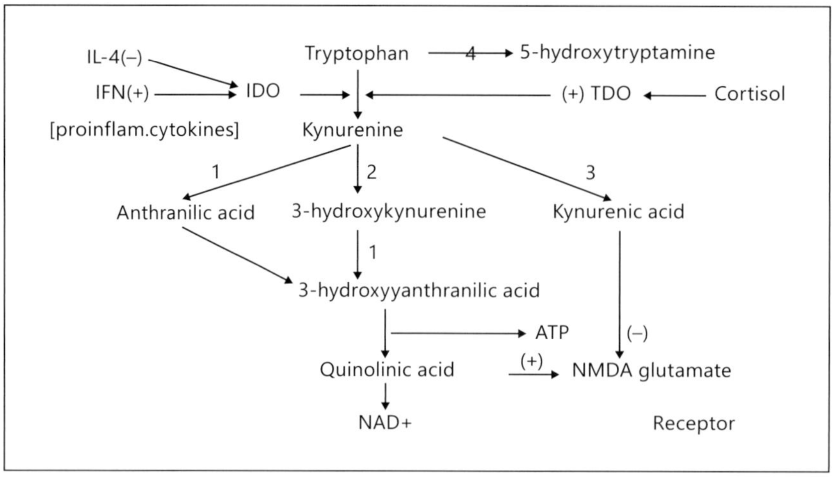

Fig. 3. The tryptophan-kynurenine pathway that links depression, pain and inflammation. 1 = Kynureninase; 2 = kynurenine oxygenase; 3 = kynurenine aminotransferase; 4 = tryptophan decarboxylase; IDO = indoleamine 2,3-dioxygenase in spleen, blood, kidney and brain; TDO = tryptophan dioxygenase in liver.

Despite the increase of the glucocorticoids in the brain and peripheral organs, the concentration of the proinflammatory cytokines is also concurrently increased in those suffering from major depression [16, 17]. This apparent contradiction is explained by the desensitisation of the glucocorticoid receptors, a situation associated with glucocorticoid resistance [18]. As a consequence, the 'feedback' inhibition of the release of the glucocorticoids, which usually arises following an acute stress response, is impaired and the secretion of the steroids from the adrenal glands continues.

The Link between Depression and Pain

It is a well-known phenomenon that the intensity of a painful experience can be modified by the emotional state. Thus, the intensity and severity of perceived pain is not usually correlated with the degree of tissue damage following trauma, and stress, fear and anxiety have important modulatory effects on pain perception [19, 20]. However, the direction of the modulation of pain by stress and emotion depends to a considerable extent on the context in which the pain occurs. In general, positive emotions inhibit the pain response regardless of the state of arousal, whereas negative emotions, even with a low-to-moderate level of arousal, enhance pain. In addition, the duration and intensity of a stressful stimulus can either reduce the pain response, a phenomenon known as stress analgesia, or enhance the pain response if the pain is chronic [21–24]. Stress also enhances existing pain associated with chronic pain disorders, including psychiatric disorders such as fibromyalgia, and inflammatory bowel disease [25, 26].

Having established that stress and the emotional state can profoundly modulate the pain response, it is not surprising to find that there is a high degree of co-morbidity between affective disorders and chronic pain. Indeed, the frequency of chronic pain in patients with major depression or anxiety has been estimated to be 22 and 35%, respectively, values which are double those of an appropriate control population [27]. As anxiety disorders are commonly associated with depression, it is not surprising to find that the frequency of chronic pain is higher than in patients suffering from depression or anxiety disorders alone [28]. Conversely, there is evidence that the frequency of depression is significantly increased in female migraine sufferers [29] and the World Health Organisation has estimated that depression is three times more frequent in those with migraine or severe headache than in healthy individuals [30]. While the pathogenesis for co-morbid chronic headache or migraine with depression is uncertain, it has been speculated that dysfunctional dopaminergic and serotonergic systems [31] and an increase in cortisol [32] are factors which link headache with depression. This accounts for stress having a role in initiating headache and depression, thereby implicating the possible involvement of inflammatory factors in a disturbance of endothelial cell function as an important component in the pathology of headache [33].

Whereas the association between inflammatory diseases such as arthritis and chronic bowel diseases is frequently overlooked, it is easily accepted that life-threatening diseases such as cancer are likely to be associated with an increase in depression. Increased pain and a progressive deterioration in the general physical health in those with advanced cancer have been estimated to increase the frequency of depression by 5–30% [34]. Of these patients, only about 25% receive appropriate antidepressant medication [35]. The possible link between pain, depression, inflammation and the therapeutic effects of antidepressants is discussed later in this review ('Can Drugs Which Decrease Inflammation Also Be Effective in the Treatment of Pain and Depression?').

In the previous section of this review, evidence was provided for the involvement of proinflammatory cytokines in major depression. TNF-α plays a critical role, particularly in the brain. Inflammatory mediators are thought to play a key role in the occurrence of hyperalgesia, and TNF-α plays a role in this process by activating the TNF receptors in the dorsal root ganglia, thereby reducing the pain threshold [36]. Evidence for the role of TNF-α in pain associated with depression is provided by the effect of a TNF antagonist injected into the dorsal root ganglia. This results in a reduction in the severity of TNF-induced pain [37]. An important clinical finding to this study is the observation that a TNF-α monoclonal antibody reduces the pain and the symptoms caused by rheumatoid arthritis [38]. Such results indicate that the block of TNF receptors is associated with a rapid reduction in the pain and disease-related symptoms even though joint swelling and tenderness do not change significantly until after the attenuation of the pain. This emphasises the dissociation between pain and other aspects of inflammation, a situation which has also been observed in the administration of the TNF-α antagonist infliximab, which improves the mood state in patients with Crohn's disease [39], an inflammatory disease that is often co-morbid with major depression.

From the foregoing discussion, it is evident that pain and depression are frequently linked with inflammation. The possible mechanisms whereby inflammation plays a critical role in this process will now be discussed.

The Association of Neuroinflammation with Pain and Depression

Many of the studies referred to in this review were based on outcome measures which depended on changes in reflexive pain, on responses that were governed by spinal cord pathways. In major depression, however, pain is associated with aversive, motivational and emotional components which rely on attention and expectation within a specific context. Some experimental studies have helped to dissect out the links between inflammatory chemokines and cannabinoid receptors in the rodent brain following spinal cord injury as a model of neuropathic pain that may be relevant to a better understanding of depression-associated pain [40].

In brief, these authors demonstrated that affective pain was correlated with an increase in type-1 cannabinoid receptors and the CCL-chemokines in the hippocampus, thalamus and periaqueductal grey areas in response to neuropathic pain [40], the chronicity of the pain being associated with the increase in the chemokines. Prior to the studies of Knerlich-Lukoschus et al. [40], Sulka [41] demonstrated that inflammation-induced pain in rats enhanced the conditioned place preference for morphine and for the N-methyl-D-aspartate (NMDA) antagonist MK-801 following the peripheral injection of the pain-inducing complete Freund's adjuvant. Other experimental studies have demonstrated that peripherally induced pain is associated with elevation of IL-1β and TNF-α in the cingulate cortex [42]. This aspect has also been assessed in a rodent study of neuropathic pain behaviour in which there was an increased expression of inflammatory genes in the amygdala [43]. The possible epigenetic effect on the enhanced inflammatory response to neuropathic pain in female rats was also shown following exposure of the animals to a stressful environment in early life [44]. Taken together, these experimental studies indicate how painful stimuli produce both brain changes and behavioural effects that also link pain with those regions that have been implicated in major depression. Furthermore, such findings suggest that inflammation is an important component of both chronic pain and chronic depression, whereby the intracellular links occur at subcellular sites beyond the cytokine receptors.

Unfortunately, the clinical studies which are necessary to substantiate the inflammation-pain-depression hypothesis are limited despite the convincing experimental data. These studies have extended the scope of the pathophysiology of pain and depression to include the purinergic system. In this regard, it has been shown that in lupus-prone mice, the anxiety and depressive-like behaviour is attenuated by suramin, a purinergic-2 receptor antagonist [45].

It is of historical interest to note that Emil Kraepelin was the first person to suggest that the purinergic system might be involved in the pathology of mood states when he observed that the excretion of uric acid, a purine, was decreased [46]. It is now apparent that the purinergic system is involved in many physiological processes which are dysfunctional in affective disorders and which are associated with changes in sleep architecture [47], anhedonia [48] and cognitive impairment [49]. The cellular link between mood disorders and the purinergic system is provided by adenosine signalling. Thus, experimentally it has been demonstrated that allopurinol and febuxostat, xanthine oxidase antagonists, have antidepressant activity which is linked to a reduction in purine synthesis [50]. Clinical studies have also shown that a dysfunctional purinergic system is associated with affective disorders [51], while a recent review of the role of the purinergic system in mood disorders has linked the purinergic-2 receptors and adenosine to both depression and inflammation [52]. Thus, the purinergic system is now recognised as playing a major role in pain, inflammation, neurodegeneration and affective disorders, and has been the subject of an extensive review [53].

While the neurochemical links between inflammation, pain and depression has focussed on the biogenic amines and the purines, a more comprehensive view of the situation has been considered recently by Walker et al. [54] by linking tryptophan metabolism to the kynurenine pathway, which is activated by proinflammatory cytokines and cortisol. Whereas there has been substantial evidence that a reduction in the free pool of tryptophan, the precursor of serotonin, is an important cause of depression [55], more recent evidence suggests that cytokine-activated indoleamine 2,3-dioxygenase and glucocorticoid activated liver tryptophan dioxygenase are also involved [56, 57]. Thus, activation of indoleamine 2,3-dioxygenase and tryptophan dioxygenase leads to the diversion of tryptophan away from serotonin synthesis through the kynurenine pathway, a situation widely occurring in the brain and in peripheral organs. The main metabolic changes that occur following tryptophan metabolism are summarised in figure 4.

A detailed discussion of the tryptophan-kynurenine pathway and its biological significance is given by Myint [58], and the clinical evidence for the increase in the neurodegenerative arm of the pathway in depressed patients has been provided by the results of the study by Myint et al. [57]. The importance of this pathway in the context of the present review arises from the evidence that proinflammatory cytokines and cortisol, which are elevated in major depression, enhance the neurodegenerative arm, which leads to an increase in the NMDA glutamate receptor activity associated with increased pain, while the neuroprotective arm, in which the NMDA receptor antagonist is decreased. Researchers have seen the accumulation of a major end product of the neurodegenerative pathway, quinolinic acid, in the microglia and other cells of the anterior cingulate cortex of depressed patients who have committed suicide [59], which provides further evidence of an increase in the neurodegenerative arm of the

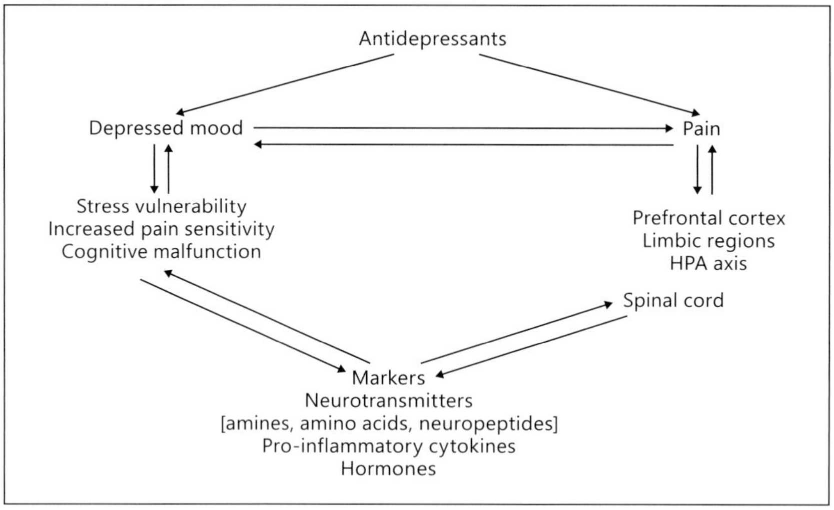

Fig. 4. The relationship between depressed mood, pain and markers of pain and depression.

tryptophan-kynurenine pathway. Clinical studies have demonstrated that an increase in glutamatergic function is positively correlated with an increase in the Hamilton Depression Score [60, 61]. Kim et al. [62] have also shown that the increase in indoleamine 2,3-dioxygenase, and presumably the consequent increase in quinolinic acid, contributes to the co-morbidity of pain and depression, which is supported by the experimental studies linking pain and depression [63].

To conclude this section of the review, evidence has been presented that chronic inflammation likely plays a crucial role that links pain with depression. At the cellular level, this is reflected in the changes in the purinergic and glutamatergic systems in addition to the monoaminergic systems which have been implicated in the depressed mood state. Of these cellular changes, the activation of the tryptophan-kynurenine pathway by inflammatory cytokines would appear to be an important pathway that enhances glutamatergic function and thereby contributes to the pathophysiological basis of pain and depression.

Can Drugs Which Decrease Inflammation Also Be Effective in the Treatment of Pain and Depression?

If chronic inflammation provides a pivotal link between pain and depression, it would be anticipated that a reduction in inflammation would have beneficial effects on these co-morbid disorders. Anecdotal reports indicate that tricyclic antidepressants such as amitriptyline have been used in the past to relieve the pain associated with chronic arthritis [64]. As the monoamine hypothesis of depression has largely driven the development of antidepressants, it is not surprising that the ability of these drugs to attenuate pain has been explained in terms of the normalisation of monoaminergic function. However, experimental studies have broadened the possible mechanisms whereby amitriptyline reduces pain, particularly neuropathic pain. One suggestion is that amitriptyline activates adenosine receptors in the spinal cord [65] and thereby postulates that the pain-gating mechanism in the dorsal horn is enhanced and the activity of sodium channels reduced. The reduction in the activity of the microglia by amitriptyline also has anti-inflammatory consequences. In a rat model of chronic depression, it has been shown that chronic amitriptyline treatment causes a more marked reduction in proinflammatory cytokines in rats with the depressive phenotype than in control rats not showing such depressive behaviour [66]. This effect of amitriptyline is probably related to its ability to selectively suppress the proinflammatory M1 state of the microglia, an effect which has been associated with a reduction in the secretion of IL-1 in the prefrontal cortex [66].

From the clinical perspective, the tricyclic antidepressants have shown efficacy in the treatment of fibromyalgia, chronic low back pain, diabetic neuropathy, postherpetic neuralgia and chronic headache, conditions in which pain and inflammation occur and which are frequently co-morbid with chronic depression [67]. However, because of the cardiotoxicity and other untoward side effects of the tricyclic antidepressants, new antidepressants with a dual action on both the noradrenergic and serotonergic transporters, such as duloxetine and venlafaxine, have achieved some prominence in the treatment of pain syndromes. In addition, the rediscovery of drugs such as the benzoxazocine compound nefopam [68], in which the antidepressant-like triple monoamine reuptake profile is combined with an inhibition of calcium flux, block of voltage-dependent sodium channels and a reduction in glutamatergic function, has opened up the possibility for developing multifunctional molecules that treat pain syndromes, depression and inflammation.

If antidepressants are therapeutically beneficial for the treatment of pain syndromes and depression, does that also mean that their primary action is the attenuation of the concurrent underlying inflammatory state? A meta-analysis of a number of clinical trials in which the effects of different classes of antidepressants on serum proinflammatory cytokines in depression were accessed demonstrated a mixed response [69]: there were consistent reductions in IL-1β but not TNF-α or IL-6. Overall it would appear that the selective serotonin reuptake inhibitor antidepressants have the most consistent profile in reducing inflammation [70], but even these are limited in reducing all the inflammatory cytokines in depression [71]. Clearly this is an important area for drug development in which the triad of pain, depression and inflammation is the target rather than the individual components. It will be of interest to see if the new generation of triple reuptake inhibitor antidepressants now in clinical development will be effective in this regard [72].

Perhaps it is time to consider drugs such as the semisynthetic tetracycline minocycline which has been used in neurology for many years and which reduces the activation of the microglia and thereby the availability of IL-1, IL-6, TNF, IL-18 and nitric oxide [73]. Clinically, minocycline reduces neuropathic pain [74] and there is experimental and 'open' trial clinical evidence that it also has antidepressant properties. Experimentally, minocycline has been demonstrated to prevent the activation and release of proinflammatory cytokines in rats following neuropathic pain by increasing the anti-inflammatory M2 state of the microglia and thereby increasing the release of the anti-inflammatory cytokine IL-10 [75].

Given the widespread evidence of the association between chronic inflammation and chronic depression, it is not surprising to find that non-steroidal anti-inflammatory drugs have been used in adjunctive therapy to treat depressed patients who have failed to respond adequately to standard antidepressant therapy. Both selective (e.g. celecoxib) and non-selective (e.g. aspirin) drugs have been investigated with mixed results [76–78]. Indeed, Shelton [74] has suggested that non-steroidal anti-inflammatory drugs may actually reduce, rather than enhance, the effectiveness of antidepressants! These anti-inflammatory drugs act by inhibiting the cyclo-oxygenases (COX) in the brain and periphery, and thereby reduce the synthesis of the inflammatory mediator prostaglandin E_2. However, the activity of the COX enzymes is complex for whereas COX-1 is predominantly proinflammatory [79, 80], COX-2 can have both beneficial and adverse effects on the brain [81]. Thus COX-2 inhibitors have been shown to increase neuroinflammation, possibly through enhanced glial activation and increased T1 cell responses [81].

It is perhaps not surprising that a recent critical systematic review of the literature for both selective and non-selective COX inhibitors as adjunctive treatment to antidepressants has concluded that the efficacy of the COX inhibitors was negligible, with the proviso that it was difficult to draw firm conclusions given the inconsistent findings and substantial differences in the heterogeneity of the patients and the methodology used [80]. It is clear that high-quality research is needed to explore the potential benefits of anti-inflammatory drugs in the adjunctive treatment of pain, depression and inflammation. The use of COX-2 inhibitors as putative antidepressants or as adjunctive treatment to antidepressants in the treatment of depression has been the subject of a critical review by Maes [82].

Conclusion

Pain as a co-morbid feature of major depression is a relatively recent consideration by psychiatrists, possibly because the treatment of pain was considered independently of any underlying psychiatric disorder. However, the more holistic approach to psychiatric problems has focussed on the importance of the interactions between 'body and mind' by emphasising the physical as well as the psychological changes that occur as a result of stress and any underlying pathology. The co-morbidity of pain and depression are well recognised in those suffering from a mental illness. Together with the recognition that chronic inflammation plays in these conditions, this has led to a better understanding of both pain and depression and how this may be of potential value in the identification of novel therapeutic targets.

The purpose of this short review was to assess the evidence, based where possible on clinical studies, in support of the psychoneuroimmune hypothesis that chronic inflammation is a key factor in the chronicity of pain and major depression. Hopefully this will eventually lead to new classes of drugs which are more effective in the control of these co-morbid conditions.

References

1 Arnow BA, Hunkeler EM, Blasey CM, et al: Comorbid depression, chronic pain and disability in primary care. Psychosom Med 2006;68:262–268.
2 Sansone RA, Sansone LA: Pain, pain go away: antidepressants and pain management. Psychiatry (Edgmont) 2008;5:16–19.
3 Han C, Pae CU: Pain and depression: a neurobiological perspective of their relationship. Psychiatry Investig 2015;12:1–8.
4 Hart BL: Biological bases of the behaviour in sick animals. Neurosci Biobehav Rev 1988;12:1232–1237.
5 Maes M, Berk M, Goehler L, et al: Depression and sickness behaviour are Janus-faced responses to shared inflammatory pathways. BMC Med 2012;10:66.
6 Kroeke K, Bair ML, Damush TM, et al: Optimized antidepressant therapy and pain self-management in primary care patients with depression and musculoskeletal pain: a randomised controlled trial. JAMA 2009;301:2099–2110.
7 Dantzer R: Cytokine-induced sickness behaviour: a neuroimmune response to activation of innate immunity. Eur J Pharmacol 2004;500:399–411.
8 Maes M, Lambrechts J, Bosmans E, et al: Evidence for a systemic immune activation during depression: results of leukocyte enumeration by flow cytometry in conjunction with monoclonal antibody staining. Psychol Med 1992;22:45–53.
9 Capuron L, Ravand A, Dantzer R: Early depressive symptoms in cancer patients receiving interleukin 2 and/or interferon alpha-2b therapy. J Clin Oncol 2000;18:2143–2151.
10 Haroon E, Raison CL, Miller AH: Psychoneuroimmunology meets neuropsychopharmacology: translational implications of the impact of inflammation on behaviour. Neuropsychopharmacology 2012;37:137–162.
11 Frank MG, Barattar MV, Springer DB, et al: Microglia serve as a neuroimmune substrate for stress-induced potentiation of CNS pro-inflammatory responses. Brain Behav Immun 2007;21:47–59.
12 Bluthe RM, Walter V, Parnet P, et al: Lipopolysaccharide induces sickness behaviour in rats mediated by vagal mediated mechanism. C R Acad Sci III 1994;317:499–503.
13 Banks WA: Blood-brain barrier transport of cytokines: a mechanism for neuropathology. Curr Pharm Des 2005;11:973–984.
14 Quan N, Banks WA: Brain-immune communication pathways. Brain Behav Immun 2007;21:727–735.
15 Pariante C, Miller AH: Glucocorticoid receptors in major depression; relevance to pathophysiology and treatment. Biol Psychiatry 2001;49:391–404.
16 Pollak Y, Yirmiya R: Cytokine induced changes in mood and behaviour; implications for 'depression due to a general medical condition', immunotherapy and antidepressant treatment. Int J Neuropsychopharmacol 2002;5:389–399.
17 Raison CL, Capuron L, Miller AH: Cytokines sing the blues: inflammation and pathogenesis of depression. Trends Immunol 2006;27:24–31.
18 Pace TW, Miller AH: Cytokines and glucocorticoid signaling. Relevance to major depression. Ann N Y Acad Sci 2009;1179:86–105.
19 Asmundson GJ, Katz J: Understanding the co-occurrence of anxiety disorders and chronic pain: state-of-the-art. Depress Anxiety 2009;26:888–901.
20 Butler RK, Finn DP: Stress-induced analgesia. Prog Neurobiol 2009;88:184–202.
21 Rudy JL, Meagher MW: Fear and anxiety; divergent effects on human pain thresholds. Pain 2000;84:65–75.
22 Crettaz B, Marziniak M, Willeke P, et al: Stress-induced allodynia; evidence of increased pain sensitivity in healthy humans and patients with chronic pain after experimentally induced social stress. PLoS One 2013;8:e69460.
23 Jennings EM, Okine BN, Roche M, Finn DP: Stress-induced hyperalgesia. Prog Neurobiol 2014;121:1–18.

24 Olango WM, Finn DP: Neurobiology of stress induced hyperalgesia. Curr Top Behav Neurosci 2014; 20:251–280.

25 Aaron LA, Buchwald D: Chronic diffuse musculoskeletal pain, fibromyalgia and co-morbid unexplained clinical conditions. Best Pract Res Clin Rheumatol 2003;17:563–574.

26 Grande LA, Loeser JD, Ozuma J, et al: Complex regional pain syndrome as a stress response. Pain 2004; 110:495–498.

27 McWilliams LA, Cox BJ, Enns MW: Mood and anxiety disorders associated with chronic pain; an examination in a nationally representative sample. Pain 2003;106:127–133.

28 Fishbain DA, Goldberg M, Meagher BR, et al: Male and female chronic pain patients categorized by DSM-3 psychiatric diagnostic criteria. Pain 1986;26:181–197.

29 Orla OR, Gelaye B, Qiu C, et al: Depression, anxiety and stress among pregnant migraineurs in a Pacific-Northwest cohort. J Affect Dis 2015;172:390–396.

30 World Health Organisation: Headache Disorders. 2012. http://www.who.int/medcentre/factsheets/fs277/er/.

31 Frediam F, Villani V: Migraine and depression. Neurol Sci 2007;28(suppl 2):S161–S165.

32 Overeem S, van Vliet JA, Lammers GJ, et al: The hypothalamus and episodic brain disorders. Lancet Neurol 2002;1:437–444.

33 Christian LM, Franco A, Glaser R, Iams JD: Depressive symptoms are associated with elevated serum proinflammatory cytokines among pregnant women. Brain Behav Immun 2009;23:750–754.

34 Mitchell AJ, Chau M, Bhatti H, et al: Prevalence of depression, anxiety and adjustment disorder in oncological, haematological and palliative care settings; a meta-analysis of 94 interview based studies. Lancet Oncol 2011;23:160–174.

35 Janberidze E, Hjermstad MJ, Brunelli C, et al; EURO IMPACT: The use of antidepressants in advanced cancer: results from an international multicentre study. Psychooncology 2014;23:1096–1102.

36 Boettger MK: Antinociceptive effects of tumour necrosis factor alpha neutralisation in a rat model of antigen-induced arthritis: evidence of a neuronal target. Arthritis Rheum 2008;58:2368–2378.

37 Marchand F: Effects of eteranercept and minaprine in a rat model of spinal cord injury. Eur J Pain 2009; 13:673–681.

38 Hess A, Axmann R, Rech J, et al: Blockade of TNF-α rapidly inhibits pain response in the central nervous system. Proc Soc Acad Sci U S A 2011;108:3731–3736.

39 Lichstenstein GR, Bala M, Han C, et al: Infliximab improves quality of life of patients with Crohn's disease. Inflamm Bowel Dis 2002;8:237–243.

40 Knerlich-Lukoschus F, Noack M, van der Ropp-Brenner B, et al: Spinal cord injuries induce changes in CB1 cannabinoid receptor and C-C chemokine expression in brain areas underlying circuitry of chronic pain conditions. J Neurotrauma 2011;28:619–634.

41 Sulka KL: Conditional place preference paradigm: a novel approach for analgesic drug assessment against chronic pain. Pain 1994;58:355–366.

42 Lu Y, Zhu L, Gao YJ: Pain-related aversion induces astrocytic reaction and proinflammatory cytokine expression in the anterior cingulated cortex in rats. Brain Res Bull 2011;84:178–182.

43 Burke NN, Geoghegan E, Kerr DM, et al: Altered neuropathic pain behaviour in a rat model of depression is associated with changes in inflammatory gene expression in the amygdala. Genes Brain Behav 2013;12:705–713.

44 Burke NN, Llorente R, Marco EM, et al: Maternal deprivation is associated with sex-dependent alterations in nociceptive behaviour and neuroinflammatory mediators in the rat following peripheral nerve injury. J Pain 2013;14:1173–1184.

45 Ballok DA, Sakic B: Purinergic receptor antagonists moderate serology and affective behaviour in lupus prone mice: evidence of autoimmune induced pain? Brain Behav Immun 2008;22:1208–1216.

46 Kraepelin E: Manic-Depressive Insanity and Paranoia. Chicago, Medical Book Co, 1921.

47 Huang ZL, Woode Y, Hayashi O: The role of adenosine in the regulation of sleep. Curr Top Med Chem 2011;11:1047–1057.

48 Csölle RA, Baranyi M, Zsille G, et al: Neurochemical changes in the mouse hippocampus underlying the antidepressant effect of genetic depletion of P2X7 receptors. PLoS One 2013;8:e66547.

49 Salomone JD, Correa M: Dopamine/adenosine interactions involved in effort-related aspects of food motivation. Appetite 2009;53:422–425.

50 Karve AV, Jagtiani SS, Chitnis KA: Evaluation of the effect of allopurinol and febuxostat in behavioural model of depression in mice. Indian J Pharmacol 2013;45:244–247, 422–425.

51 Bauer A, Ishiwata K: Adenosine receptor ligands and PET imaging in the CNS. Handb Exp Pharmacol 2009;617–642.

52 Ortiz R, Ulrich H, Zarate CA, Machado-Vieira R: Purinergic system dysfunction in mood disorders: a key target for developing improved therapeutics. Prog Neuropsychopharm Biol Psychiatry 2015;57:117–131.

53 Burnstock G: Physiolgy and pathophysiology of purinergic transmission. Physiol Rev 2007;87:659–797.

54 Walker AK, Kavelaars A, Heijnen CJ, Dantzer R: Neuroinflammation and comorbidity of pain and depression. Pharmacol Rev 2014;66:80–101.

55 Cowen PL, Perry-Billings M, Newsholme FA: Decreased plasma tryptophan levels in major depression. J Affect Dis 1989;16:27–31.

56 Wichers MC, Koek GH, Robaeys G, et al: Indoleamine deoxygenase and interferon-alpha induced degenerative symptoms; a shift in hypothesis from tryptophan depletion to neurotoxicity. Mol Psychiatry 2000;10:538–544.
57 Myint AM, Kim YK, Verkerk R, et al: Kynurenine pathway in major depression: evidence of impaired neuroprotection. J Affect Disord 2007;98:143–151.
58 Myint AM: Inflammation, neurotoxins and psychiatric disorder. Modern Trends Pharmacopsychiatri 2013;28:61–74.
59 Steiner J, Walter M, Gos T, et al: Severe depression is associated with increased microglia quinolinic acid in subregions of the anterior cingulate gyrus: evidence for an immune-modulated glutamatergic neurotransmission. J Neuroinflammation 2011;8:94.
60 Kim JS, Schmid-Burgk W, Claus D, Kornhuber HH: Increased serum glutamate in depressed patients. Arch Psychiatr Nervenkr (1970) 1982;232:299–304.
61 Mitani H, Shirayama Y, Yamada T, et al: Correlation between plasma levels of glutamate, alanine and serine with severity of depression. Prog Neuropsychopharmacol Biol Psychiatry 2006;30:1155–1158.
62 Kim H, Chen L, Lim G, et al: Brain indoleamine 2,3-dioxygenase contributes to the comorbidity of pain and depression. J Clin Invest 2012;122:2940–2954.
63 Li JX: Pain and depression comorbidity: a preclinical perspective. Behav Brain Res 2015;276:92–98.
64 Mico JA, Ardidy D, Berrocoso E, et al: Antidepressants and pain. Trends Pharmacol Sci 2006;27:348–354.
65 Esser MJ, Chase T, Allen GV, Sawynok J: Chronic administration of amitryptyline and caffeine in a rat model of neuropathic pain; multiple interactions. Eur J Pharmacol 2012;430:211–218.
66 Burke NN, Finn DP, Roche M: Chronic administration of amitriptyline differentially alters neuropathic pain-related behaviour in the presence and absence of depressive-like phenotype. Behav Brain Res 2015; 278:193–201.
67 Chan HN, Fan J, Ng BY: Use of antidepressants in the treatment of chronic pain. Ann Acad Med Singapore 2009;38:974–979.
68 Kim KH, Abdi S: Rediscovery of nefopam for the treatment of neuropathic. Korean J Pain 2014;27: 103–111.
69 Hannestad J, DellaGioia N, Bloch M: The effect of antidepressant medication treatment on serum levels of inflammatory cytokines: a meta-analysis. Neuropsychopharmacology 2011;36:2452–2459.
70 Basterzi AD, Aydemir C, Kisa C, et al: IL-6 levels decrease with SSRI treatment in patients with major depression. Hum Psychopharmacol 2005;20:473–476.
71 O'Brien SM, Scully P, Fitzgerald P, et al: Plasma cytokine profiles of depressed patients who fail to respond to selective serotonin reuptake inhibitor therapy. J Psychiatr Res 2007;41:326–331.
72 Marks DM, Pac CU, Patkar AA: Triple reuptake inhibitors: a premise and a promise. Psychiatry Investig 2008;5:142–147.
73 Rojewska E, Popiolek-Barczyk K, Jurga AM, et al: Involvement of pro- and antinociceptive factors in minocycline analgesia in rat neuropathic pain model. J Neuroimmunol 2014;277:57–66.
74 Shelton RC: Does concomitant use of NSAIDs reduce the effectiveness of antidepressants? Am J Psychiatry 2012;169:1012–1015.
75 Burke NN, Kerr DM, Moriarty O, et al: Minocycline modulates neuropathic pain behaviour and cortical M1-M2 microglia gene expression in a rat model of depression. Brain Behav Immun 2014;42:147–156.
76 Akhondzadeh S, Jafari S, Raisi F, et al: Clinical trial of adjunctive celecoxib treatment in patients with major depression: a double-blind and placebo controlled trial. Depress Anxiety 2009;26:607–611.
77 Almeida OP, Flicker L, Yeap BB, et al: Aspirin decreases the risk of depression in old men with high plasma homocysteine. Transl Psychiatry 2012;2: e151.
78 Müller N, Schwarz MJ, Dehning S, et al: The cyclooxygenase-2-inhibitor celecoxib has therapeutic effects in depression; results of a double-blind, randomised, placebo controlled, add-on pilot study with reboxetine. Mol Psychiatry 2006;11:680–684.
79 Aid S, Bose HF: Targeting cyclooxygenase-1 and-2 in neuroinflammation, therapeutic implications. Biochimie 2011;93:46–51.
80 Berk M, Dean O, Drexhage H, et al: Aspirin: a review of its neurobiological properties and therapeutic potential for mental illness. BMC Med 2013;11:74.
81 Eyre HA, Air T, Proctor S, et al: A critical review of the efficacy of non-steroidal anti-inflammatory drugs in depression. Prog Neuropsychopharmacol Biol Psychiatry 2015;57:11–16.
82 Maes M: Targeting cyclooxygenase-2 in depression is not a viable therapeutic approach and may even aggravate the pathophysiology underpinning depression. Metab Brain Dis 2012;27:405–413.

Prof. Brian E. Leonard
Faculty of Science, Department of Pharmacology
University College Galway
Galway (Ireland)
E-Mail psycholeonard@googlemail.com

Finn DP, Leonard BE (eds): Pain in Psychiatric Disorders. Mod Trends Pharmacopsychiatry.
Basel, Karger, 2015, vol 30, pp 36–50 (DOI: 10.1159/000435931)

Neuroinflammatory Mechanisms Linking Pain and Depression

Nikita N. Burke[a–c] • David P. Finn[b, c] • Michelle Roche[a, c]

[a]Physiology, [b]Pharmacology and Therapeutics, School of Medicine, and [c]Galway Neuroscience Centre and Centre for Pain Research, National University of Ireland, Galway, Ireland

Abstract

Depression and chronic pain have been estimated to co-occur in up to 80% of patients suffering from these disorders, with this co-morbidity being more disabling and more expensive to both patients and society than either disorder alone. A number of neural substrates have been proposed to underlie this association; however, there has been increased interest and support for a role of neuroimmune and neuroinflammatory mechanisms as key players in this dyad. This chapter will provide an overview of the clinical and preclinical data supporting a role for neuroimmune alterations in depression-pain co-morbidity. We propose that such changes may impact on the functioning of key brain regions modulating emotional and nociceptive processing, thus resulting in the behavioural, psychological and physical symptoms observed in patients exhibiting depression and co-morbid pain.

Introduction to Depression-Pain Co-Morbidity

Depression is one of the most prevalent forms of psychiatric illness and is associated with significant disability, impaired health-related quality of life and high mortality [1, 2]. Depressed patients display a constellation of psychological (depressed mood, anhedonia, feelings of worthlessness or guilt, altered concentration, and suicidality) and physical (altered appetite and weight, disturbed sleep, psychomotor agitation or retardation, and fatigue) symptoms [3]. In addition to being a debilitating disorder in its own right, depression is frequently associated with other disorders including neurological conditions, cancer, cardiovascular disease and chronic pain. For example, depression and chronic pain have been estimated to co-occur in up to 80% of patients [4], and the combination is more disabling and costlier to both patients and society than either disorder alone [5]. Thus, there is an urgent need to understand the underlying mechanisms facilitating depression-pain co-morbidity in order to ensure appropriate treatment and facilitate the development of novel treatments. A number of potential common neural substrates have been implicated in the co-morbidity of these conditions, including neurotransmitters, neurotrophins, inflammatory mediators and neuroendocrine alterations (for a review, see Maletic and Raison [6]), any or all of which may alter neural functioning in key brain regions involved in regulating emotional and nociceptive processing. While ascending and descending pain pathways are critical in the somatosensory aspects of nociception, there is a well-recognised role for cortical and limbic regions in modulating these nociceptive pathways and regulating emotional, cognitive and affective responding to painful stimuli [7–9].

Detailed discussion of the various mediators which may alter activity of supraspinal areas modulating emotional and nociceptive processing is beyond the scope of this chapter and has been discussed in detail in several excellent reviews [6, 10, 11]. An abundance of evidence supports a role for altered inflammatory processes in the pathophysiology of chronic pain and psychiatric disorders independently. Increasingly, glial and neuroimmune processes are considered possible mediators in the association between these disorders. This chapter will examine the evidence for a role of neuroinflammation in the depression-pain dyad.

Peripheral Immune Alterations Observed in Depression-Pain Co-Morbidity

As highlighted, a wealth of evidence supports a role for immune processes in the pathophysiology of depression and chronic pain individually, and although it may be intuitive that alterations would also be observed in co-morbid patients, few studies have examined whether chronic pain patients with depression, or vice versa depressed patients with comorbid pain, exhibit alterations in circulating immune mediators. One recent study examining associations between pain thresholds and peripheral levels of inflammatory mediators in outpatients with major depressive disorder demonstrated elevated plasma levels of the proinflammatory cytokine TNF-α and reduced pressure pain thresholds in these patients [12]. Correlative analysis revealed that augmented levels of TNF-α correlated significantly with reduced pain thresholds in depressed women, but not men [12]. To our knowledge, this is the only study to date that has examined and shown sexual dimorphic effects between immune mediators, depression and pain responding. Further support for alterations in inflammatory mediators in depression-pain co-occurrence arises from a primary care setting study where increased plasma levels of the proinflammatory cytokine

IL-6 were associated with greater reported pain in patients exhibiting depressive symptoms, an effect not observed in patients without depressive symptomatology [13]. In addition, patients with chronic back pain and co-morbid depression have been shown to exhibit higher plasma IL-6 levels compared to healthy controls [14]. Similarly, depressive symptoms and serum IL-6 levels correlated with pain in patients with burning mouth syndrome [15] and enhanced neuropathic pain was associated with higher depression symptoms and serum C-reactive protein levels in sciatica patients [16]. A recent study demonstrated that in female fibromyalgia patients, depressive symptoms positively correlate with elevated plasma C-reactive protein levels [17]. In addition, chronic lower back pain patients have been reported to have high serum levels of TNF-α; and although the presence of co-morbid depression did not influence TNF-α levels, pain patients with depression benefited more from multidisciplinary treatment than patients without depression [18]. Thus, taken together, the data indicate that depression-pain co-morbidity is associated with enhanced peripheral proinflammatory cytokine levels, effects which may underlie (at least in part) the pathophysiology of this association.

Neuroinflammation Associated with Depression-Pain Co-Morbidity

While inflammatory mediators may modulate nociceptive processing at a peripheral level, communication with the central nervous system is required for alterations in affective and emotional processing. Cytokines communicate with the brain via several routes including circumventricular organs, saturable blood-brain barrier transporters and activation of the vagus nerve (for a review, see Raison et al. [19]). This immune-to-brain signalling influences neurotransmitter metabolism, the HPA axis and neuroendocrine function, synaptic plasticity, and neural circuits, and results in behavioural changes such as depressed mood, fatigue, cognitive dysfunction and hyperalgesia. Glial activation and the subsequent release of proinflammatory cytokines is usually temporary, restrained by anti-inflammatory mechanisms, and is an adaptive survival response. However, chronic and sustained neuroinflammation may result in pathological changes within the brain and negatively impact on neuronal circuitries. As such, chronic neuroinflammation in key brain regions mediating emotional and nociceptive processing may underlie depression-pain co-morbidity.

While post-mortem studies have indicated region-specific microglial activation and increased cytokine expression in pain- and mood-related brain regions including the thalamus and the prefrontal cortex (PFC) in subjects previously diagnosed with depression [20–23], it is unknown if any of these patients also exhibited altered nociceptive responding or a chronic pain condition. Conversely, to date, post-mortem studies examining neuroinflammatory alterations associated with chronic pain have concentrated on changes within the spinal cord, demonstrating significant activation of both microglia and/or astrocytes and increased inflammatory cytokine expression in these patients [24, 25]. While post-mortem data are useful, a number of confounds can influence the data interpretation. The advent of newer imaging technologies has enabled a more detailed examination of real-time neuroinflammatory processes at a structural and functional level. For example, a recent review highlighted the use of magnetic resonance spectroscopy to assess neuroinflammation in painful conditions including low back pain, neuropathy, migraine and fibromyalgia [26]. Using this technology, researchers have found reduced N-acetylaspartate levels (suggesting neuronal injury or loss) and increased myo-inositol (glial activation) in regions such as the frontal cortex and thalamus in a number of chronic pain conditions [26], regions which have also shown neuroinflammation in post-mortem studies of depressed patients. Positron emission tomography (PET) and magnetic resonance spectroscopy have revealed increased glial activation in the thalamus of patients with limb denervation [27], and in the PFC [28] and thalamus [29] of spinal cord injury patients with neuropathic pain, regions which also play a role in modulating mood. To date, only one study has used this technology to examine microglia activity in a depressed cohort, and although alterations were not detected using currently available PET ligands [30], a number of post-mortem histological and molecular studies [24–27] have indicated that enhanced neuroinflammation is associated with this disorder.

Using this non-invasive approach, a recent study reported the first evidence for supraspinal neuroinflammation in patients exhibiting co-morbid depression and pain symptoms [31]. In this study, microglial activation was assessed in patients with myalgic encephalitis/chronic fatigue syndrome who exhibited pain and depression using PET assessment of the radiotracer ^{11}C-(R)-PK11195. Increased binding of the radiotracer, and thus microglial activation, in the hippocampus was positively correlated with depression scores while binding in the thalamus tended to be correlated positively with pain scores in these patients, indicating that region-specific neuroinflammation may be associated with symptoms of depression and pain. It should be noted that while patients exhibiting depression-pain symptomatology exhibited robust microglial activation, only mild peripheral inflammation was observed in these patients, i.e. a non-significant tendency for enhanced interferon-γ levels. Thus, these data indicate that neuroinflammation can occur in the absence of overt peripheral inflammation, and may underlie the changes in emotional and nociceptive processing observed in depression-pain comorbidity. Due to the heterogeneity of depressive and chronic pain disorders, the fact that inflammatory markers vary dynamically over time, and the confounding factors such as treatment or disease stage, it is difficult to achieve a clear consensus on the role of inflammation in the co-morbidity of these conditions using only clinical studies. Therefore, appropriate preclinical models are essential in order to examine the neurobiological mechanisms underpinning the link between mood and pain disorders and identify new treatment strategies.

Preclinical Evidence Supporting a Role for Inflammation in Emotion and Pain Interactions

Studies in laboratory animals support clinical data on the relationship between depression and pain, and provide a means to investigate the underlying mechanisms and neural substrates mediating the association between these disorders. As a wealth of data demonstrates neuroimmune changes in models of depression and chronic pain independently (for a review, see Song and Wang [32] and Austin and Moalem-Taylor [33]), that animal models of chronic pain display depressive- and anxiety-like behaviour [34] and that several animal models of depression exhibit altered nociceptive responding [35], the remainder of this chapter will provide an overview of the data supporting the hypothesis that neuroimmune processes are involved in the interaction between these conditions.

Animal Models of Chronic Pain Exhibit Depressive-Like Behaviour and Neuroimmune Activation

It is now well-established that depressive-like behaviour can be observed in a wide variety of preclinical chronic pain models. Rodent models of neuropathic and inflammatory pain have been shown to display depressive-like behaviours such as increased immobility in the forced swim test, anhedonia in the sucrose preference test and deficits in the social interaction test (for a review, see Yalcin et al. [34]). Ample evidence supports a role for immune processes and glial activation in the spinal cord in mediating enhanced nociceptive processing in chronic pain (for a review, see Vallejo et al. [36] and Bradesi [37]) and it has been shown that spinal microglia activation is critical in the development of neuropathic pain, while in comparison, astrocyte activation is essential for the maintenance of this pain state [38]. Furthermore, evidence indicates that chronic pain in rodents results in glial activation and cytokine and chemokine release in discrete regions including the cingulum, hippocampus and PFC [39–42], effects which may have implications for the descending control of pain, affective pain behaviour and emotional processing. In support of this, recent studies indicate that a neuroinflammatory response occurs in discrete brain regions following chronic inflammatory and neuropathic pain, which is associated with the development of depressive-like behaviour.

In the complete Freund's adjuvant model of inflammatory pain, both rats and mice have been shown to exhibit concomitant depressive-like behaviour and increased cytokine expression in the brain, specifically increased levels of the proinflammatory cytokine IL-1β in the whole brain [43] and increased IL-6 mRNA in the hippocampus [14]. Thus, inflammatory pain-induced depressive-like behaviour is associated with neuroinflammation, a consequence that is not unsurprising, given the immune cascade that occurs in such a model. While this supports a possible association between neuroinflammation, pain and depression, the precise role of the individual cytokines within the brain mediating the nociceptive and/or emotional behavioural changes in the model remain to be determined.

Peripheral nerve injury (spared nerve injury) in mice induces a pronounced mechanical allodynia which is accompanied by the development of depressive-like behaviour as determined by enhanced immobility in the forced swim test, increased IL-1β expression in the frontal cortex and increased astrocyte activation in the periaqueductal grey, a key brain region for the descending modulation of pain [44]. Blocking central IL-1β using the IL-1 receptor antagonist attenuated pain-induced depressive-like behaviour, demonstrating a causative role of IL-1β in nerve injury-induced depression. Interestingly, chronic stress or social isolation exacerbated spared nerve injury-induced mechanical allodynia and depressive-like behaviour, and resulted in a greater enhancement of IL-1β mRNA in the frontal cortex, indicating that stressful events may potentiate pain and precipitate depression by promoting neuroinflammation [44, 45]. Extending these findings, Zhou et al. [46] showed that spared nerve injury-induced mechanical hyperalgesia and depressive-like behaviour was accompanied by increased IL-1β mRNA in the liver, ipsilateral lumbar dorsal horn and contralateral frontal cortex, highlighting concomitant peripheral, spinal and supraspinal neuroinflammation in a model exhibiting depression and pain behaviour.

The chronic constriction injury model of neuropathic pain has been shown to exhibit anhedonia (a core depressive behaviour) and impaired hippocampal neurogenesis/plasticity in a time-dependent manner. Furthermore, TNF protein levels were increased and TNF receptor subtype 2 (TNFR2) expression, the neuroprotective receptor subtype, was decreased in the hippocampus. The aforementioned behavioural and neurogenesis effects were not seen in TNFR1 knockout mice, indicating that neuropathic pain-related depression and hippocampal plasticity are dependent on TNF signalling through TNFR1 [47].

Taken together, these studies show that the development of depressive-like behaviour following peripheral nerve injury is accompanied by enhanced IL-1β and TNF signalling in the frontal cortex and hippocampus, respectively, suggesting that neuroimmune alterations induced by persistent nociceptive input to brain regions that process mood may account, at least in part, for behavioural deficits such as depression. Moreover, these studies demonstrate a causal role of these proinflammatory cytokines in neuropathic pain-induced depressive-like behaviour.

Animal Models of Depression Exhibit Altered Nociceptive Thresholds – Possible Role for Inflammatory Mediators?

While depressive-like behaviour has been demonstrated in a wide variety of animal models of chronic pain (for a review, see Yalcin et al. [34]), until relatively recently, less was known about whether animal models of depression exhibited altered nociceptive responding. However, several studies have now demonstrated that a variety of well-validated models of depression based on stress, lesion, genetics and pharmacological manipulation display altered nociceptive responding, mimicking the clinical picture. For example, the Wistar-Kyoto (WKY) rat, a stress hyperresponsive rat strain with a depressive-like phenotype, exhibits thermal hyperalgesia in the hot plate test

[48], visceral hyperalgesia to colorectal distension [49–51] and hyperalgesia to intraplantar injection of formalin [48]. The reserpine-induced depression model results in pronounced and long-lasting mechanical hyperalgesia and allodynia, and cold allodynia [52, 53]. The chronic mild stress model has been shown to display cold allodynia [54, 55], mechanical allodynia and hyperalgesia [54, 56], and thermal hypoalgesia [57, 58]. Early-life stress is known to result in depressive-like behaviour in adulthood and results in altered sensitivity to thermal stimuli [59–62], increased sensitivity to mechanical stimuli [62, 63] and visceral hypersensitivity [50, 51, 59, 64]. Recent work from our laboratory and others has shown that the olfactory bulbectomised (OB) rat, a lesion model of depression, exhibits increased sensitivity to mechanical and thermal stimuli in the von Frey, acetone drop, hot plate and tail flick tests [48, 65–68] while exhibiting reduced thermal and mechanical sensitivity in the paw pressure and Hargreaves tests [69–71], indicating effects which may be dependent on the type, intensity and duration of stimuli applied. Thus, a variety of animal models of depression exhibit altered nociceptive thresholds, mimicking the changes in sensitivity to noxious stimuli as seen in depressed patients.

Animal models of depression are well known to exhibit altered peripheral and central inflammatory processing. For example, early-life stress results in increased astrocyte density in the hippocampus and cerebellum [72–74]. WKY rats exhibit increased astrocyte expression in the dorsal raphe nuclei [75] and reductions in astrocyte expression in the PFC and amygdala [76]. Microglial activation and increased protein levels of proinflammatory cytokines have been reported following chronic/social stress in the PFC, periaqueductal grey, amygdala and hippocampus [77–79], and OB rats exhibit increased IL-1β, TNF-α, prostaglandins and astrocyte activation in the brain [80–83]. Therefore, models of depression exhibit region-specific alterations in glial activation in sites implicated in emotional and pain processing, effects which may contribute to the altered nociceptive responding in these models of depression.

In addition to examining nociceptive thresholds, research has also examined if models of depression exhibit altered nociceptive responding in a chronic persistent pain state. An abundance of data now shows that the induction of inflammatory pain consistently results in enhanced nociceptive responding across a range of models of depression. Specifically, inflammatory nociceptive responding has been reported to be enhanced in the WKY rat [48, 84, 85], the OB rat [48, 57, 69], following chronic stress [14, 54, 86, 87] and in the early-life stress [61] models of depression (table 1). Several of these studies have demonstrated altered levels of monoamines in key brain sites in models of depression following inflammatory pain, indicating that the inflammatory stimulus results in alterations in neuronal functioning that may underlie the hyperalgesia observed. For example, research from our group has demonstrated that both OB and WKY rats exhibit enhanced nociceptive behaviour in response to intraplantar administration of formalin, effects correlating with alterations in the monoamine levels in key brain sites regulating emotional and nociceptive processing [48]. It is possible that the heightened inflammatory processes observed in models of

Table 1. Preclinical animal studies demonstrating relationship between inflammation-depression and pain

Chronic pain model	Depressive-like behaviour	Inflammatory changes/effects	
Kaolin and carrageenan-induced monoarthritis	↓ Sucrose preference ↑ Immobility in FST	Not measured	[96]
CFA-induced monoarthritis	↑ Immobility in FST	Not measured	[97]
CFA-evoked inflammatory pain	↓ Sucrose preference ↑ Immobility in FST and TST	↑ IL-1β levels in whole brain, and cortex ↑ Plasma IL-6, ↑ IL-6 in hippocampus	[43] [14]
Spinal nerve injury (neuropathic)	↑ Immobility in FST ↓ Social interaction	↑ IL-1β mRNA in the frontal cortex ↑ GFAP mRNA in the periaqueductal grey ↑ IL-1β mRNA liver ↑ IL-1β mRNA dorsal horn of spinal cord	[44–46]
Chronic constriction injury (neuropathic)	↓ Sucrose preference	↑ TNF and ↓ TNFR2 in hippocampus	[47]
Model of depression	**Nociceptive responding/behaviour**	**Inflammatory changes/effects**	
Reserpine-induced depression model	Mechanical allodynia in von Frey test Mechanical hyperalgesia in paw pressure test Thermal hyperalgesia in tail immersion test	↑ TNF-α, IL-1β and substance P in cortex and hippocampus	[88–90]
OB rat	Mechanical allodynia in von Frey test Cold allodynia in acetone drop test	↑ GFAP (astrocyte activation), CD11b (microglial activation) and IL-1β mRNA in the amygdala	[66]
	↑ Formalin-evoked nociceptive behaviour	Not measured	[48, 69]
	↑ Nociceptive responding to electrical stimulation of the dura mater (model of trigeminal nociception)	↑ Plasma substance P and CGRP levels	[98]
	Bilateral mechanical allodynia following SNL Increased cold allodynia following SNL	↓ IL-6 and IL-10 mRNA in amygdala, ↑ Microglia (CD68, CD40, CD11b) and astrocyte (GFAP) activation, anti-inflammatory (IL-10) cytokines and proinflammatory chemokines (CCL2, CCL5) in mRNA in PFC	[65, 66]
WKY rat	↑ Formalin-evoked nociceptive behaviour ↑ Mechanical hyperalgesia following CFA to the TMJ	Not measured	[48, 85] [84]
	↑ Mechanical allodynia following nerve injury	↑ IL-1β mRNA in the brainstem and PFC; ↑ IL-1β and IL-6 in hippocampus	[40, 92, 99]

Table 1. Continued

Model of depression	Nociceptive responding/behaviour	Inflammatory changes/effects	
Unpredictable chronic mild stress	↑ Nociceptive responding to electrical stimulation of the dura mater	↑ Plasma substance P levels	[100]
Maternal deprivation	↑ Mechanical and cold allodynia following SNL in females only	↓ IL-6 mRNA in PFC; ↑ GFAP and IL-1β mRNA in hippocampus of MD-SNL males ↓ TNF-α mRNA in PFC; ↑ in IL-6 and TNF-α mRNA in hippocampus of MD-SNL females	[62]
Maternal separation	↑ Formalin-evoked nociceptive behaviour	Not measured	[61]
Chronic restraint stress	↑ Formalin-evoked nociceptive behaviour	Not measured	[54, 86]

CCL2 = Chemokine C-C motif ligand 2; CCL5 = chemokine C-C motif ligand 5; CD11b = cluster of differentiation 11b; CD40 = cluster of differentiation 40; CD68 = cluster of differentiation 68; CFA = complete Freund's adjuvant; CGRP = calcitonin gene-related peptide; FST = forced swim test; GFAP = glial fibrillary acidic protein; MD = maternal deprivation; mRNA = messenger RNA; TMJ = temporomandibular joint; TST = tail swim test.

depression may prime the system such that when challenged with an exogenous inflammatory stimulus, pain behaviour is enhanced. Few studies have examined this directly; however, rats subjected to chronic social defeat exhibit spinal neuroinflammation, specifically increased iNOS and COX-2 mRNA in the dorsal lumbar spinal cord, which was associated with mechanical hyperalgesia [87]. Additionally, repeated reserpine treatment results in increased TNF-α and IL-1β in the cortex and hippocampus of rats and mice, effects accompanied by increased nociceptive responding and depressive-like behaviour [88–90]. Thus, while peripheral mechanisms cannot be ruled out, it is possible that neuroinflammatory processing may underlie, at least in part, the altered nociceptive thresholds observed in rodent models of depression.

Animal Models of Depression Exhibit Altered Neuropathic Pain Responding and Accompanying Neuroimmune Alterations

A number of studies have examined whether the development and expression of neuropathic pain is altered in animal models of depression (table 1). Pain-related behaviour is altered in the majority of these studies; however, the effects vary depending on the model of depression and means of neuropathic pain induction. A thorough overview of all the data in this area is beyond the scope of this chapter; however, presented herein are studies that support a role for neuroinflammation in the altered behavioural responding in these models.

Recent data from our laboratory reported that following L5-L6 spinal nerve ligation (SNL), the OB rat model of depression exhibits bilateral mechanical allodynia and altered nociceptive responding to an innocuous cold stimulus when compared to sham controls [66, 67], an effect accompanied by bilateral astrocyte activation in the dorsal horn of the spinal cord (unpublished). Spinal astrocytes have been implicated in the development of mirror-image mechanical allodynia following nerve injury [91]; therefore, these data suggest that astrocyte activation at the level of the spinal cord may contribute to the bilateral mechanical allodynia observed in OB rats following nerve injury. Interestingly, supraspinal neuroinflammation is also associated with altered pain responding in the OB-SNL rat. For instance, OB rats exhibited reduced latency and number of responses to an innocuous cold stimulus following SNL, an effect positively correlated with IL-6 and IL-10 mRNA expression in the amygdala [66]. Moreover, OB-SNL rats exhibited heightened expression of markers of microglia and astrocyte activation, anti-inflammatory cytokines and proinflammatory chemokines in the PFC when compared to either OB or SNL alone [65, 67]. The functional role of these inflammatory mediators in the PFC on affective and nociceptive behaviour in the present model remains to be determined, yet these data clearly indicate that the combination of depression and pain leads to altered inflammatory gene expression in critical brain regions.

The WKY rat model of depression also exhibits exacerbated mechanical allodynia following peripheral nerve injury (chronic constriction injury) compared to non-depressed Wistar controls [92]. IL-1β mRNA expression was increased in the brainstem and PFC of WKY rats following peripheral nerve injury, and IL-1β levels in the PFC correlated with mechanical sensitivity [40]. Thus, the proinflammatory profile in the PFC may underlie the heightened neuropathic pain-related behaviour associated with the depressive-like phenotype of WKY rats.

Gender is an important and often overlooked factor in research – females are twice as likely to suffer from depression as males, have greater pain sensitivity and are more likely to experience chronic pain [93, 94]. We have recently shown that following peripheral nerve injury (SNL), females, but not males, exposed to early-life stress exhibited exacerbated nerve injury-induced mechanical and cold allodynia compared with non-stressed counterparts [62]. This behavioural change was accompanied by reduced TNF-α in the PFC with a concomitant increase in IL-6 and TNF-α expression in the hippocampus, compared with either maternal deprivation or SNL alone. As such, the combination of early-life stress and peripheral nerve injury resulted in sexually dimorphic alterations in neuropathic pain-related behaviour and the expression of genes encoding inflammatory mediators depending on the brain region. Sex-dependent changes in neuroimmune functioning in these regions may underlie the higher incidence of depression and pain disorders in women.

Taken together, these studies show that the induction of neuropathic pain in an animal model with a pre-existing depressive-like phenotype results in exacerbated pain-related responding accompanied by differential inflammatory gene expression in brain regions responsible for the regulation of emotion and pain. These data corroborate limited clinical data that show increased depressive symptoms in patients with neuropathic pain are associated with increased inflammation, at least at a peripheral level and expand it to provide evidence for supraspinal inflammation. Further studies are required to determine the precise mechanisms by which microglia and astrocytes modulate neuronal activity and influence emotional and nociceptive processing.

Modulating (Neuro)Inflammation Reduces Depression-Like and Pain-Related Behaviour in Preclinical Models

If neuroinflammatory mechanisms truly underlie the co-occurrence of depression and pain, then modulation of such processes should be of therapeutic benefit to patients. Studies have examined effects of systemic, intrathecal and central administration of anti-inflammatory compounds on emotional and nociceptive processing. For example, oral treatment with resveratrol, an anti-inflammatory antioxidant, has been shown to reduce depressive-like behaviour and thermal hyperalgesia in a model of neuropathic pain (chronic constriction injury), effects which were abolished by depletion of central serotonin [95], highlighting a possible interaction between inflammation and monoamine neurotransmission that may underlie the association between depression and pain. Indeed, while antidepressants are known to modulate monoaminergic neurotransmission, these pharmacological agents are also known to exhibit anti-inflammatory effects, and amitriptyline, a tricyclic antidepressant, is a first-line treatment for neuropathic pain. We have recently reported that chronic systemic amitriptyline administration to OB rats that have undergone nerve injury (SNL), results in an attenuation of depressive-like behaviour and SNL-induced mechanical allodynia [65]. These behavioural changes were accompanied by an attenuation of astrocyte activation, the expression of the anti-inflammatory cytokine IL-10 and the chemokine CCL5, and enhanced the expression of TNF-α in the PFC. Thus, chronic antidepressant treatment can alleviate depressive-like and neuropathic pain responding possibly by normalising neuroinflammatory processes in brain regions that influence emotional and nociceptive responding such as the PFC.

Further evidence supporting a role for neuroimmune processing in altered neuropathic pain responding associated with a depressive phenotype arises from a recent study where we demonstrated that chronic administration of minocycline, a microglial inhibitor, attenuated OB-induced depressive-like behaviour and prevented the development of SNL-induced mechanical and cold allodynia in OB rats. Strikingly, chronic minocycline treatment enhanced the anti-inflammatory M2 microglial phenotype in the PFC of OB-SNL animals in contrast to sham-SNL animals where minocycline rather modulated the M1 proinflammatory phenotype. Evidence suggests that the phenotype of activated microglia (M1 vs. M2) governs the repair and regeneration response following nerve injury and these data suggest a differential mechanism of action of minocycline in the presence of a depressive-like phenotype, which may have important implications for clinical therapy.

Conclusion and Future Directions

The current data indicate that regardless of the primary disorder, the combination of depression and pain is associated with dysregulated inflammation in the central nervous system. The behavioural outputs may represent molecular, cellular, chemical, structural and functional alterations in the brain triggered by neuroinflammatory changes in specific brain areas responsible for parallel processing of emotion and nociception. However, further clinical studies are warranted to confirm and extend the preclinical data, and the precise mechanisms and sites by which neuroimmune processes influence emotion and nociception remain to be determined.

Acknowledgements

This work was funded by grants from Science Foundation Ireland Research Frontiers Project (11/RFP/NES/3175), the Discipline of Physiology and a PhD Scholarship from the College of Medicine, National University of Ireland Galway.

References

1 Ormel J, et al: Common mental disorders and disability across cultures. Results from the WHO Collaborative Study on Psychological Problems in General Health Care. JAMA 1994;272:1741–1748.
2 Spitzer RL, et al: Health-related quality of life in primary care patients with mental disorders. Results from the PRIME-MD 1000 Study. JAMA 1995;274:1511–1517.
3 DSM-5, Diagnostic and Statistical Manual of Mental Disorders, ed 5. Arlington, American Psychiatric Publishing, 2013.
4 Poole H, et al: Depression in chronic pain patients: prevalence and measurement. Pain Pract 2009;9:173–180.
5 Gameroff MJ, Olfson M: Major depressive disorder, somatic pain, and health care costs in an urban primary care practice. J Clin Psychiatry 2006;67:1232–1239.
6 Maletic V, Raison CL: Neurobiology of depression, fibromyalgia and neuropathic pain. Front Biosci (Landmark Ed) 2009;14:5291–5338.
7 Graff-Guerrero A, et al: Cerebral blood flow changes associated with experimental pain stimulation in patients with major depression. J Affect Disord 2008;107:161–168.
8 Lopez-Sola M, et al: Effects of duloxetine treatment on brain response to painful stimulation in major depressive disorder. Neuropsychopharmacology 2010;35:2305–2317.
9 Bar KJ, et al: Increased prefrontal activation during pain perception in major depression. Biol Psychiatry 2007;62:1281–1287.
10 Walker AK, et al: Neuroinflammation and comorbidity of pain and depression. Pharmacol Rev 2014;66:80–101.
11 Bair MJ, et al: Depression and pain comorbidity: a literature review. Arch Intern Med 2003;163:2433–2445.
12 Euteneuer F, et al: Depression, cytokines and experimental pain: evidence for sex-related association patterns. J Affect Disord 2010;131:143–149.
13 Poleshuck EL, et al: Depressive symptoms, pain, chronic medical morbidity, and interleukin-6 among primary care patients. Pain Med 2013;14:686–691.
14 Kim H, et al: Brain indoleamine 2,3-dioxygenase contributes to the comorbidity of pain and depression. J Clin Invest 2012;122:2940–2954.
15 Chen Q, et al: Serum interleukin-6 in patients with burning mouth syndrome and relationship with depression and perceived pain. Mediators Inflamm 2007;2007:45327.
16 Uher T, Bob P: Neuropathic pain, depressive symptoms, and C-reactive protein in sciatica patients. Int J Neurosci 2013;123:204–208.
17 Menzies V, et al: Psychoneuroimmunological relationships in women with fibromyalgia. Biol Res Nurs 2013;15:219–225.

18 Wang H, et al: Influence of comorbidity with depression on interdisciplinary therapy: outcomes in patients with chronic low back pain. Arthritis Res Ther 2010;12:R185.
19 Raison CL, Capuron L, Miller AH: Cytokines sing the blues: inflammation and the pathogenesis of depression. Trends Immunol 2006;27:24–31.
20 Steiner J, et al: Immunological aspects in the neurobiology of suicide: elevated microglial density in schizophrenia and depression is associated with suicide. J Psychiatr Res 2008;42:151–157.
21 Shelton RC, et al: Altered expression of genes involved in inflammation and apoptosis in frontal cortex in major depression. Mol Psychiatry 2011;16: 751–762.
22 Pandey GN, et al: Proinflammatory cytokines in the prefrontal cortex of teenage suicide victims. J Psychiatr Res 2012;46:57–63.
23 Tonelli LH, et al: Elevated cytokine expression in the orbitofrontal cortex of victims of suicide. Acta Psychiatr Scand 2008;117:198–206.
24 Del Valle L, Schwartzman RJ, Alexander G: Spinal cord histopathological alterations in a patient with longstanding complex regional pain syndrome. Brain Behav Immun 2009;23:85–91.
25 Shi Y, et al: Chronic-pain-associated astrocytic reaction in the spinal cord dorsal horn of human immunodeficiency virus-infected patients. J Neurosci 2012;32:10833–10840.
26 Chang L, et al: Magnetic resonance spectroscopy to assess neuroinflammation and neuropathic pain. J Neuroimmune Pharmacol 2013;8:576–593.
27 Banati RB, et al: Long-term trans-synaptic glial responses in the human thalamus after peripheral nerve injury. Neuroreport 2001;12:3439–3442.
28 Widerstrom-Noga E, et al: Metabolite concentrations in the anterior cingulate cortex predict high neuropathic pain impact after spinal cord injury. Pain 2013;154:204–212.
29 Pattany PM, et al: Proton magnetic resonance spectroscopy of the thalamus in patients with chronic neuropathic pain after spinal cord injury. AJNR Am J Neuroradiol 2002;23:901–905.
30 Hannestad J, et al: The neuroinflammation marker translocator protein is not elevated in individuals with mild-to-moderate depression: a [^{11}C]PBR28 PET study. Brain Behav Immun 2013;33:131–138.
31 Nakatomi Y, et al: Neuroinflammation in patients with chronic fatigue syndrome/myalgic encephalomyelitis: an ^{11}C-(R)-PK11195 PET Study. J Nucl Med 2014;55:945–950.
32 Song C, Wang H: Cytokines mediated inflammation and decreased neurogenesis in animal models of depression. Prog Neuropsychopharmacol Biol Psychiatry 2011;35:760–768.
33 Austin PJ, Moalem-Taylor G: The neuro-immune balance in neuropathic pain: involvement of inflammatory immune cells, immune-like glial cells and cytokines. J Neuroimmunol 2010;229:26–50.
34 Yalcin I, Barthas F, Barrot M: Emotional consequences of neuropathic pain: insight from preclinical studies. Neurosci Biobehav Rev 2014;47C:154–164.
35 Li JX: Pain and depression comorbidity: a preclinical perspective. Behav Brain Res 2014.
36 Vallejo R, et al: The role of glia and the immune system in the development and maintenance of neuropathic pain. Pain Pract 2010;10:167–184.
37 Bradesi S: Role of spinal cord glia in the central processing of peripheral pain perception. Neurogastroenterol Motil 2010;22:499–511.
38 Raghavendra V, Tanga F, DeLeo JA: Inhibition of microglial activation attenuates the development but not existing hypersensitivity in a rat model of neuropathy. J Pharmacol Exp Ther 2003;306:624–630.
39 Zhu J, et al: Interaction of glia activation and neurotransmission in hippocampus of neuropathic rats treated with mirtazapine. Exp Clin Psychopharmacol 2009;17:198–203.
40 Apkarian AV, et al: Expression of IL-1beta in supraspinal brain regions in rats with neuropathic pain. Neurosci Lett 2006;407:176–181.
41 Ignatowski TA, et al: Brain-derived TNFalpha mediates neuropathic pain. Brain Res 1999;841:70–77.
42 Wei F, et al: Supraspinal glial-neuronal interactions contribute to descending pain facilitation. J Neurosci 2008;28:10482–10495.
43 Maciel IS, et al: Synergistic effects of celecoxib and bupropion in a model of chronic inflammation-related depression in mice. PLoS One 2013;8:e77227.
44 Norman GJ, et al: Stress and IL-1beta contribute to the development of depressive-like behavior following peripheral nerve injury. Mol Psychiatry 2010;15: 404–414.
45 Norman GJ, et al: Social interaction prevents the development of depressive-like behavior post nerve injury in mice: a potential role for oxytocin. Psychosom Med 2010;72:519–526.
46 Zhou W, et al: 118. Indoleamine 2,3 dioxygenase (IDO1) is a key regulator of depressive-like behavior but not of mechanical hyperalgesia in the spared nerve injury model of neuropathic pain. Brain Behav Immun 2013;32:e34.
47 Dellarole A, et al: Neuropathic pain-induced depressive-like behavior and hippocampal neurogenesis and plasticity are dependent on TNFR1 signaling. Brain Behav Immun 2014;41:65–81.
48 Burke NN, et al: Enhanced nociceptive responding in two rat models of depression is associated with alterations in monoamine levels in discrete brain regions. Neuroscience 2010;171:1300–1313.

49 Gibney SM, et al: Colorectal distension-induced prefrontal cortex activation in the Wistar-Kyoto rat: implications for irritable bowel syndrome. Neuroscience 2009;165:675–683.
50 Gosselin RD, et al: Riluzole normalizes early-life stress-induced visceral hypersensitivity in rats: role of spinal glutamate reuptake mechanisms. Gastroenterology 2010;138:2418–2425.
51 O'Malley D, et al: Distinct alterations in colonic morphology and physiology in two rat models of enhanced stress-induced anxiety and depression-like behaviour. Stress 2010;13:114–122.
52 Ogino S, et al: Systemic administration of 5-HT(2C) receptor agonists attenuates muscular hyperalgesia in reserpine-induced myalgia model. Pharmacol Biochem Behav 2013;108:8–15.
53 Nagakura Y, et al: Different pathophysiology underlying animal models of fibromyalgia and neuropathic pain: comparison of reserpine-induced myalgia and chronic constriction injury rats. Behav Brain Res 2012;226:242–249.
54 Bardin L, et al: Chronic restraint stress induces mechanical and cold allodynia, and enhances inflammatory pain in rat: relevance to human stress-associated painful pathologies. Behav Brain Res 2009; 205:360–366.
55 Bravo L, et al: Depressive-like states heighten the aversion to painful stimuli in a rat model of comorbid chronic pain and depression. Anesthesiology 2012;117:613–625.
56 Imbe H, et al: Chronic restraint stress decreases glial fibrillary acidic protein and glutamate transporter in the periaqueductal gray matter. Neuroscience 2012; 223:209–218.
57 Shi M, Wang JY, Luo F: Depression shows divergent effects on evoked and spontaneous pain behaviors in rats. J Pain 2010;11:219–229.
58 Qi Q, et al: Depressive-like history alters persistent pain behavior in rats: opposite contribution of frontal cortex and amygdala implied. Psych J 2013;2:133–145.
59 Coutinho SV, et al: Neonatal maternal separation alters stress-induced responses to viscerosomatic nociceptive stimuli in rat. Am J Physiol Gastrointest Liver Physiol 2002;282:G307–G316.
60 Weaver SA, Diorio J, Meaney MJ: Maternal separation leads to persistent reductions in pain sensitivity in female rats. J Pain 2007;8:962–969.
61 Uhelski ML, Fuchs PN: Maternal separation stress leads to enhanced emotional responses to noxious stimuli in adult rats. Behav Brain Res 2010;212:208–212.
62 Burke NN, et al: Maternal deprivation is associated with sex-dependent alterations in nociceptive behavior and neuroinflammatory mediators in the rat following peripheral nerve injury. J Pain 2013;14:1173–1184.
63 Alvarez P, Green PG, Levine JD: Stress in the adult rat exacerbates muscle pain induced by early-life stress. Biol Psychiatry 2013;74:688–695.
64 Moloney RD, et al: Early-life stress induces visceral hypersensitivity in mice. Neurosci Lett 2012;512:99–102.
65 Burke NN, Finn DP, Roche M: Chronic administration of amitriptyline differentially alters neuropathic pain-related behaviour in the presence and absence of a depressive-like phenotype. Behav Brain Res 2015;278:193–201.
66 Burke NN, et al: Altered neuropathic pain behaviour in a rat model of depression is associated with changes in inflammatory gene expression in the amygdala. Genes Brain Behav 2013;12:705–713.
67 Burke NN, et al: Minocycline modulates neuropathic pain behaviour and cortical M1-M2 microglial gene expression in a rat model of depression. Brain Behav Immun 2014;42:147–156.
68 Rodriguez-Gaztelumendi A, et al: An altered spinal serotonergic system contributes to increased thermal nociception in an animal model of depression. Exp Brain Res 2014;232:1793–1803.
69 Wang W, et al: The differential effects of depression on evoked and spontaneous pain behaviors in olfactory bulbectomized rats. Neurosci Lett 2010;472: 143–147.
70 Belcheva I, et al: Differential involvement of hippocampal vasoactive intestinal peptide in nociception of rats with a model of depression. Peptides 2009;30: 1497–1501.
71 Su YL, et al: The effect of depression on the thermal nociceptive thresholds in rats with spontaneous pain. Neurosci Bull 2010;26:429–436.
72 Llorente R, et al: Early maternal deprivation in rats induces gender-dependent effects on developing hippocampal and cerebellar cells. Int J Dev Neurosci 2009;27:233–241.
73 Marco EM, et al: Maternal deprivation effects on brain plasticity and recognition memory in adolescent male and female rats. Neuropharmacology 2013;68:223–231.
74 Lopez-Gallardo M, et al: Maternal deprivation and adolescent cannabinoid exposure impact hippocampal astrocytes, CB1 receptors and brain-derived neurotrophic factor in a sexually dimorphic fashion. Neuroscience 2012;204:90–103.
75 Pearson KA, et al: Identifying genes in monoamine nuclei that may determine stress vulnerability and depressive behavior in Wistar-Kyoto rats. Neuropsychopharmacology 2006;31:2449–2461.
76 Gosselin RD, et al: Region specific decrease in glial fibrillary acidic protein immunoreactivity in the brain of a rat model of depression. Neuroscience 2009;159:915–925.

77 Farooq RK, et al: Is unpredictable chronic mild stress (UCMS) a reliable model to study depression-induced neuroinflammation? Behav Brain Res 2012; 231:130–137.
78 Tynan RJ, et al: Chronic stress alters the density and morphology of microglia in a subset of stress-responsive brain regions. Brain Behav Immun 2010;24: 1058–1068.
79 Mormede C, et al: Chronic mild stress in mice decreases peripheral cytokine and increases central cytokine expression independently of IL-10 regulation of the cytokine network. Neuroimmunomodulation 2002;10:359–366.
80 Rinwa P, Kumar A, Garg S: Suppression of neuroinflammatory and apoptotic signaling cascade by curcumin alone and in combination with piperine in rat model of olfactory bulbectomy induced depression. PLoS One 2013;8:e61052.
81 Borre Y, et al: Minocycline restores spatial but not fear memory in olfactory bulbectomized rats. Eur J Pharmacol 2012;697:59–64.
82 Song C, Zhang XY, Manku M: Increased phospholipase A2 activity and inflammatory response but decreased nerve growth factor expression in the olfactory bulbectomized rat model of depression: effects of chronic ethyl-eicosapentaenoate treatment. J Neurosci 2009;29:14–22.
83 Cizkova D, Racekova E, Vanicky I: The expression of B-50/GAP-43 and GFAP after bilateral olfactory bulbectomy in rats. Physiol Res 1997;46:487–495.
84 Wang S, et al: Exacerbated mechanical hyperalgesia in rats with genetically predisposed depressive behavior: role of melatonin and NMDA receptors. Pain 2012;153:2448–2457.
85 Rea K, et al: Impaired endocannabinoid signalling in the rostral ventromedial medulla underpins genotype-dependent hyper-responsivity to noxious stimuli. Pain 2014;155:69–79.
86 Gameiro GH, et al: The effects of restraint stress on nociceptive responses induced by formalin injected in rat's TMJ. Pharmacol Biochem Behav 2005;82: 338–344.
87 Rivat C, et al: Chronic stress induces transient spinal neuroinflammation, triggering sensory hypersensitivity and long-lasting anxiety-induced hyperalgesia. Pain 2010;150:358–368.
88 Arora V, et al: Curcumin ameliorates reserpine-induced pain-depression dyad: behavioural, biochemical, neurochemical and molecular evidences. Psychoneuroendocrinology 2011;36:1570–1581.
89 Xu Y, et al: Ferulic acid increases pain threshold and ameliorates depression-like behaviors in reserpine-treated mice: behavioral and neurobiological analyses. Metab Brain Dis 2013;28:571–583.
90 Arora V, Chopra K: Possible involvement of oxido-nitrosative stress induced neuro-inflammatory cascade and monoaminergic pathway: underpinning the correlation between nociceptive and depressive behaviour in a rodent model. J Affect Disord 2013; 151:1041–1052.
91 Obata H, et al: Activation of astrocytes in the spinal cord contributes to the development of bilateral allodynia after peripheral nerve injury in rats. Brain Res 2010;1363:72–80.
92 Zeng Q, et al: Exacerbated mechanical allodynia in rats with depression-like behavior. Brain Res 2008; 1200:27–38.
93 Tsang A, et al: Common chronic pain conditions in developed and developing countries: gender and age differences and comorbidity with depression-anxiety disorders. J Pain 2008;9:883–891.
94 Munce SE, Stewart DE: Gender differences in depression and chronic pain conditions in a national epidemiologic survey. Psychosomatics 2007;48: 394–399.
95 Zhao X, et al: Chronic resveratrol treatment exerts antihyperalgesic effect and corrects co-morbid depressive like behaviors in mice with mononeuropathy: involvement of serotonergic system. Neuropharmacology 2014;85:131–141.
96 Amorim D, et al: Amitriptyline reverses hyperalgesia and improves associated mood-like disorders in a model of experimental monoarthritis. Behav Brain Res 2014;265:12–21.
97 Borges G, et al: Reversal of monoarthritis-induced affective disorders by diclofenac in rats. Anesthesiology 2014;120:1476–1490.
98 Liang J, et al: The effects of OB-induced depression on nociceptive behaviors induced by electrical stimulation of the dura mater surrounding the superior sagittal sinus. Brain Res 2011;1424:9–19.
99 del Rey A, et al: Chronic neuropathic pain-like behavior correlates with IL-1beta expression and disrupts cytokine interactions in the hippocampus. Pain 2011;152:2827–2835.
100 Zhang M, et al: Effects of UCMS-induced depression on nociceptive behaviors induced by electrical stimulation of the dura mater. Neurosci Lett 2013; 551:1–6.

Dr. Michelle Roche
Physiology, School of Medicine, National University of Ireland
University Road
Galway (Ireland)
E-Mail michelle.roche@nuigalway.ie

Finn DP, Leonard BE (eds): Pain in Psychiatric Disorders. Mod Trends Pharmacopsychiatry.
Basel, Karger, 2015, vol 30, pp 51–66 (DOI: 10.1159/000435932)

Cytokines in Neuropathic Pain and Associated Depression

Justin G. Lees · Brett Fivelman · Samuel S. Duffy · Preet G.S. Makker · Chamini J. Perera · Gila Moalem-Taylor

School of Medical Sciences, The University of New South Wales, Sydney, N.S.W., Australia

Abstract

Neuropathic pain occurs as a result of lesion or disease affecting the somatosensory nervous system and is present in a diverse set of peripheral and central pathologies such as nerve trauma, diabetic neuropathy, post-herpetic neuralgia, chemotherapy-induced peripheral neuropathy, spinal cord injury and multiple sclerosis. Debilitating symptoms including allodynia, hyperalgesia and spontaneous pain have a substantial negative impact on patients' quality of life. The currently available therapeutic treatments are generally ineffective and characterised by poor response rates. Accumulating evidence suggests that neuroinflammation and cytokine signalling play a critical role in neuropathic pain. Numerous experimental studies have demonstrated that certain pro-inflammatory cytokines are elevated in neuropathic pain conditions, and administration of these cytokines can elicit pain hypersensitivity in the absence of injury or disease. This phenomenon is also apparent in the 'sickness response', which encompasses a broad inflammatory response to disease and injury and involves a series of physiological and behavioural changes including pain hypersensitivity. Interestingly, the 'sickness response' is also similar in nature to some of the defining characteristics of the depressed state of affective disorder. In this review, we explore links that may relate the co-existence of depression in neuropathic pain patients with the activity of cytokines and discuss the role of several key pro-inflammatory and anti-inflammatory cytokines in neuropathic pain.

Neuropathic Pain

Nociceptive pain is a protective adaptive phenomenon that alerts the body to noxious stimuli. However, damage resulting in lesion or disease to the somatosensory nervous system can cause neuropathic pain [1], a maladaptive phenomenon characterised by spontaneous ongoing or shooting pain and amplified stimulus-evoked pain [2, 3]. The diagnosis of neuropathic pain can be complicated by the presence of alternative forms of pain (acute or inflammatory) and inherent difficulties in determining the precise anatomical location of the primary causative somatosensory disease or lesion [4–6]. Additionally, substantial variability in methodologies used for reporting and assessment and the diversity of underlying causes make it difficult to accurately report prevalence [1]. Despite these limitations, neuropathic pain is known to affect a significant proportion of the general population with estimates ranging from approximately 3 to 8% [7, 8].

A heterogeneous set of conditions are linked with the development of neuropathic pain, most commonly metabolic disorders (e.g. diabetic peripheral neuropathy), infection-related neuropathy (e.g. post-herpetic neuralgia), autoimmune disorders of the nervous system (e.g. multiple sclerosis), pharmacological treatment (e.g. chemotherapy-induced peripheral neuropathy) and traumatic nervous system damage (e.g. spinal cord injury). Symptoms associated with neuropathic pain are well described and can be similar across the different conditions. Depending on the location and severity of damage, patients may suffer from any or all of the following: allodynia (perception of pain from innocuous stimuli), hyperalgesia (heightened perception of pain from noxious stimuli), paraesthesia and dysesthesia (abnormal sensations such as pins and needles, tingling, burning and electric-shock sensations). In patients, neuropathic pain is usually chronic and frequently experienced as spontaneous pain, which occurs in the absence of stimulus. These characteristics contribute substantially to the burdensomeness and adverse quality of life reported to accompany this condition [9, 10]. In addition, currently available drugs inadequately treat neuropathic pain. There are several frontline therapeutic options including various anti-depressants, anti-convulsive drugs and opioids; however, these are characterised by poor responder rates and bothersome side effects [11, 12]. Thus, despite some success, there is an outstanding need to develop better treatment options for neuropathic pain and primarily to gain a better understanding of the complex signalling processes involved in the development and prolongation of this condition.

In recent decades, development of animal models has aided research investigating the aetiology of neuropathic pain [13, 14]. Initially, animal models focused on peripheral nerve injury in rodents, and several variations of sciatic nerve injury have been described. For example, chronic constriction injury (CCI) of the sciatic nerve reliably induces mechanical allodynia and thermal hyperalgesia in the hind paws of both mice and rats [13]. Subsequently, animal models were developed for investigation of other conditions associated with clinical neuropathic pain including spinal cord injury, chemotherapy-induced peripheral neuropathy, diabetic neuropathy and virally induced neuropathy, among others. In all these animal models, pain hypersensitivity can be reliably measured using various reflexive techniques. A major criticism of animal behavioural testing for neuropathic pain is the lack of effective methods for the assessment of spontaneous pain [15]. Attempts have being made to develop new techniques to address this issue, including a facial grimace scale for the quantification of acute pain via facial expressions [16] and conditioned place preference for measuring ongoing pain [17].

Neuropathic Pain and Neuroinflammation

While several causative mechanisms have been implicated in the development of neuropathic pain, including repetitious ectopic firing of affected nerves, peripheral sensitisation, central sensitisation in the spinal cord and abnormal processing of pain signalling in higher brain areas, it is now clear that in all cases a neuroimmune response is triggered by the underlying neuronal damage [18]. In certain conditions, the immune system has been implicated in the primary initiation of the neuropathy (e.g. Guillain-Barre syndrome, multiple sclerosis), and in other conditions there is a secondary immune response that addresses the neuronal damage [19].

Peripheral nerve damage prompts resident immunosurveillant mast cells, macrophages and myelin-producing Schwann cells to coordinate an inflammatory response aimed at defending against invading pathogens and repairing neuronal tissue [20, 21]. These activated resident immune cells secrete matrix metalloproteases and vasoactive mediators, thereby increasing the permeability of local blood vessels, and express chemokine attractant molecules that recruit circulating neutrophils [22], monocyte-derived macrophages and lymphocytes [23, 24]. The newly arrived cells are activated by cytokines, critically important signalling molecules of the immune system, produced in abundance by immune cells at the site of inflammation. Activated phagocytic macrophages serve to clear debris [21, 25] in a process of Wallerian degeneration, whereby damaged nerve fibres are degraded, replaced and remyelinated [26]. During this process, immune cells and damaged neurons release inflammatory mediators including adenosine triphosphate [23], histamine [27], substance P [28] and calcitonin gene-related peptide [29] that can sensitise nociceptors, thereby contributing to the neuropathic pain condition. Furthermore, cytokines produced during this process further sensitise nociceptors by directly signalling to neurons expressing cytokine receptors [30, 31].

Continued inflammation leads to an adaptive immune response and the infiltration of T cells at multiple levels of the neuroaxis [32–34]. T cells are demonstrated to contribute to the development and progression of neuropathic pain in animal models, particularly by the secretion of pro-inflammatory cytokines. Yet, it is important to recognise that T cells expressing anti-inflammatory cytokines also alleviate pain hypersensitivity resulting from peripheral nerve injury by helping to resolve inflammation and restore homeostasis [32]. Therefore, cytokine secretion by both innate and adaptive immune cells plays an important role in regulating the inflammatory environment and modulating the development and progression of neuropathic pain.

Additionally, following peripheral nerve injury, glial cells of the dorsal root ganglia (DRG) and spinal cord (satellite glial cells; microglia and astrocytes) become activated and also participate in the production of cytokines and sensitisation of nociceptive neurons. In circumstances where the injury occurs within the central nervous system, glial cells play a pivotal role in the immune response to injury [19, 35] and further the progression of neuropathic pain.

Neuropathic Pain, Depression and Cytokines

Chronic pain conditions, including pain classified as neuropathic [10], are often accompanied by depression [36]. A survey of 7,707 Austrians found that 260 (3.3%) fitted the criteria of suffering from neuropathic pain, and 34% of these reported suffering from depression [37]. Co-existence of depression in chronic pain patients correlates with poorer physical and psychosocial functioning [38], whilst low levels of depression in outpatients consistently predict improved function and reduced pain [39]. There is also evidence to suggest that pain-related disability and the severity and permanency of depressive symptoms are correlated [40]. Patients with diabetic peripheral neuropathic pain and co-existent depression accessed medical care and received pain relieving medication at significantly higher levels when compared to those without co-existent depression [41]. Similarly, in a 6-month follow-up study of patients with herpes zoster, onset diagnosis revealed that those patients suffering from post-herpetic neuralgia presented with higher scores not only for pain but also depression when compared to those who did not develop neuralgia [42]. Various other studies indicate a correlation between the incidence of neuropathic pain and depression, but the nature of the causal relationship is far from clear. This raises the question of whether there are biochemical linkages that might provoke a state of depression in neuropathic pain patients.

Neuropathic pain patients often report suffering from certain classic symptoms of depression such as fatigue, lethargy and weight fluctuation. These symptoms are also characteristics of what has been called the ‘sickness response’: a term coined to describe the adaptive phenomenon whereby animals assume a protective mode, aiming to conserve energy and focus on healing [43]. The ‘sickness response’ has been demonstrated to include a component of hyperalgesia, presumably as an incentive to reduce activity and act with precaution. ‘Sickness response’ hyperalgesia is apparent after intraperitoneal injection of endotoxins such as lipopolysaccharide. The response is mediated by pro-inflammatory cytokines and can be elicited in the absence of immune challenge by direct intraperitoneal injection of pro-inflammatory cytokines [44]. ‘Sickness response’ hyperalgesia is induced by peripheral cytokines directly signalling to centrally located neurons through active transport across the blood-brain barrier, infiltration at weak points such as the circumventricular structures, transmitting signals via brain endothelial cells or via vagus and glossopharyngeal sensory nerves [45]. Depression is also characterised by both elevated pro-inflammatory cytokine expression [46, 47] and by many of the ailments included in the ‘sickness response’. It has been suggested that in conditions such as neuropathic pain, the ‘sickness response’ is exacerbated and becomes maladaptive [45], a suggestion that could also apply for depression.

There is a significant overlap in the areas of the brain that are functionally or structurally abnormal in patients with major depressive disorder and those suffering from neuropathic pain, including regions of the prefrontal cortex, the anterior cingulate cortex, insula, hippocampus, thalamus and amygdala [48]. Depression is characterised by hypothalamic-pituitary-adrenal axis dysfunction, abnormalities in glucocorticoid signalling [49] and the turnover of monoamine neurotransmitters [50] and metabolism of glutamate, substance P and brain-derived neurotropic factor [51]. Many of these signalling abnormalities are likewise associated with neuropathic pain [52]. Furthermore, a growing body of evidence suggests that cytokines, critical to the neuroimmune response in neuropathic pain, also play an important role in the development of depression [53]. This idea is particularly well illustrated by findings demonstrating that patients treated with interferon (IFN)-α for either hepatitis C [54] or malignant melanoma [55] developed treatment-related depression. As outlined in the previous section, cytokines are crucially involved in the process of nerve repair and neuroinflammation, and are capable of eliciting neuronal hypersensitivity either directly or indirectly. In the context of depression, numerous animal model studies have demonstrated a link between abnormal cytokine signalling and the induction of depression-like characteristics [56]. In humans, a meta-analysis of combined studies measuring cytokine expression in depressed and non-depressed patients found a significant elevation in tumour necrosis factor (TNF)-α and interleukin (IL)-6 in depressed patients. Other conspicuous pro-inflammatory cytokines such as IL-1β and IFN-γ, were not found to be significantly elevated in depressed patients. However, there was substantial variation for these cytokines, with some studies indicating dramatic increases whilst others suggested large decreases, a finding that may relate to the cross-sectional design of the studies assessed [46]. A separate meta-analysis study of cytokine expression in humans indicated that IL-6 and IL-1 were positively associated with depression; however, the effect size associated with IL-1 was relatively mild [47].

A suggested biochemical linkage between neuropathic pain, depression and cytokine production is through the serotonin metabolic pathway. Neuropathic pain results in an inflammatory response that involves increased production of cytokines. Cytokines such as TNF, IL-6 and IFN-γ induce the activity of indoleamine 2,3-dioxygenase (IDO), thereby reducing the plasma concentration of tryptophan and increasing the production of TRYCATs (tryptophan catabolites along the IDO pathway). TRYCATs such as kynurenine and quinolinic acid have been shown to induce a depressive state due to neurotoxic effects [57, 58]. Several other biochemical linkages are apparent including the cytokine-induced dysfunction of the hypothalamic-pituitary-adrenal axis [59], which regulates aspects of both neuropathic pain and depression.

Due to these biochemical linkages, it is now plausible that cytokine signalling resulting from neuropathic pain could elicit a bystander effect on the condition of depression. In the following section, we review the involvement in neuropathic pain of key pro-inflammatory cytokines TNF, IL-1β, IFN-γ and IL-6, which have all also frequently been associated with the condition of depression [46, 47, 53, 58]. Furthermore, we also review the involvement in neuropathic pain of IL-17, a more recently discovered pro-inflammatory cytokine, and the anti-inflammatory cytokines TGF-β and IL-10, all of which are potentially associated with the condition of depression.

The Role of Individual Cytokines in Neuropathic Pain

Increasingly well-defined categorisation of cytokines has been developed based upon functional properties. The broadest classification places cytokines in either pro-inflammatory or anti-inflammatory groups. Pro-inflammatory cytokines act in the initial stages of immune response and repair, and then in normal circumstances during the resolution stage, anti-inflammatory cytokines are produced in order to depress immune activity, reduce inflammation and restore homeostasis. Here, we individually review the main cytokines relevant to neuropathic pain.

Pro-Inflammatory Cytokines

Tumour Necrosis Factor

TNF is a key mediator of pro-inflammatory responses related to necrosis, apoptosis and proliferation. It belongs to a superfamily of ligand/receptor proteins called the TNF/TNF receptor (TNFR) superfamily proteins [60]. TNF is produced by a variety of cell types, including immune cells, glial cells and neurons in response to inflammatory stimuli. It exists in two forms, both of which are biologically active [61]; soluble TNF is a homotrimer of 17-kDA monomers, cleaved from transmembrane TNF, a homotrimer of 26-kDA uncleaved monomers. Following its enzymatic cleavage from its cell-bound precursor, soluble TNF is released from cells and exerts biological effects via binding to its receptors TNFR1 and TNFR2 [62]. Whilst TNFR1 is strongly activated by both forms of TNF, TNFR2 signalling is only efficiently activated by transmembrane TNF [62, 63].

TNF plays an important role in the generation and maintenance of neuropathic pain. Pro-inflammatory cytokine expression levels reported from human nerve biopsies and serum analyses suggest that TNF is elevated in patients with painful neuropathies compared to non-painful neuropathies [64, 65]. In animal models of neuropathic pain, TNF and TNF receptors are upregulated at the site of injury, primarily in Schwann cells and macrophages [66, 67]. Intrasciatic administration of TNF elicits CCI-like pain hypersensitivity [67, 68], which is reversed by antagonism of TNF by etanercept, a recombinant TNFR fusion protein [69] or inhibition of TNF synthesis by thalidomide [70]. Similarly, in mice lacking TNF or TNFR1, neutrophil infiltration is decreased at the site of injury following peripheral nerve injury. These animals also exhibit reduced mechanical allodynia, which is restored to levels comparable to wild-type mice following administration of intraneural recombinant TNF [71]. It has been proposed that TNF mediates central mechanisms of neuropathic pain via glial activation [72, 73]. TNF is upregulated in spinal microglia and astrocytes of mice following nerve injury [73, 74], and suppression of the glial response reduces pain hypersensitivity [75]. Furthermore, following CCI in rats, TNF is upregulated in the hippocampus, locus coeruleus and red nucleus [76]. Mice with CCI-induced neuropathic pain demonstrate both depressive symptoms and decreased hippocampal neurogenesis. These changes coincide with hippocampal upregulation of TNF, mediated via TNFR1 signalling [77].

Interleukin-1β

The IL-1 cytokine superfamily consists of 11 members, including the archetypal IL-1α, IL-1β and receptor antagonist IL-1RA. IL-1β is a pleiotropic cytokine known to be involved in immune and inflammatory responses occurring in association with a wide variety of disease states including pathogen-based infections, physical injury, cancer and autoimmune disorders [78]. IL-1β signals through the IL-1 receptor type I (IL-1R1) present on the surface of numerous assorted target cell types, and produces signalling downstream to both the MAPK and NFκβ pathways, resulting in substantial global gene regulation [79]. Neurons within many areas of the central nervous system are demonstrated to express the IL-1R1, including the amygdala [80], which is a key pain and stress processing centre.

The involvement of IL-1β in nociceptive pain signalling has been repeatedly demonstrated. Hind paw injury due to complete Freud's adjuvant injection leads to an intense increase in IL-1β production at the site of injury, which remains elevated for >5 days [81]. Significant elevation of IL-1β expression in the lumbar spinal cord is also detected in rats after CCI and cryoneurolysis of the sciatic nerve [82]. Similarly, acute inflammation via hind paw injection of zymosan resulted in elevated IL-1β expression in the lumbar spinal cord [83]. Administration of IL-1β injected both into the hind paw and intraperitoneally induced hyperalgesia [84] and increased the pain response associated with subcutaneous formalin injection [85]. Central injection into the lateral cerebral ventricle induced thermal hyperalgesia; however, development of hyperalgesia was found to be dose dependent [86]. Intrathecal injection of IL-1β in normal rats produced neuronal excitability accompanied by a transient increase in mechanical allodynia [87]. Intrathecal injection also resulted in robust thermal hyperalgesia lasting up to 24 h that was associated with inducible nitric oxide synthase (iNOS) activity [88]. iNOS is a key contributor to mechanisms involved in the perception of nociceptive pain [89]. IL-1β was further proposed to act upon the p38 MAPK, and it was found that downregulation of this pathway resulted in inhibition of the associated thermal hyperalgesia and iNOS activity [90]. p38 MAPK is expressed exclusively in microglia and not in astrocytes or neurons in the spinal cord following peripheral nerve injury, suggesting that microglia are crucial in the initiation of IL-1β neuropathic pain signalling [91]. Knockout of the adenosine triphosphate ligand-gated ion channel P2X7 decreased IL-1β expression and abolished hyperalgesia and allodynia associated with nerve injury. This suggests a model whereby neuronal damage initiates pulsatile release of adenosine triphosphate, which acts upon P2X7 receptors expressed predominantly by local immune cells. P2X7 signalling induces the production and release of IL-1β from immune cells [92]. IL-1β can also directly activate nociceptor signalling via the IL-1R1, which is expressed on neurons in the DRG, activating p38 MAPK and directly contributing to the hypersensitivity of neurons [93]. Both genetic impairment of IL-1 signalling and chronic blockade of IL-1 attenuated neuropathic pain symptoms and spontaneous ectopic neuronal activity following peripheral nerve injury in mice [94, 95].

Interleukin-6

IL-6 is a pleiotropic cytokine produced by a wide range of hematopoietic and somatic cells, and it influences many cell types with multiple biological activities. IL-6 belongs to a family of neuropoietic cytokines and is thought to be a major factor in the response to a nervous system injury [96]. Accumulating evidence suggests that IL-6 is involved in a diverse range of functions including neuroprotection, promotion of neuronal survival, induction of neuronal differentiation, modulation of biosynthesis of neurotransmitters, stimulation of astrocyte proliferation and modulation of pain. IL-6 first binds to membrane-specific IL-6 receptor (IL-6R/gp80), which is then combined with its transmembranous signal transducer gp130, resulting in a cascade of intracellular phosphorylation of several signal proteins [97, 98]. Two major intracellular signalling pathways for IL-6 have been documented: the classical pathway acts through Janus kinases and STAT factors, whereas the alternative pathway uses the Ras-dependent MAPK cascade, which is associated with intracellular signal proteins involved in nociceptive potentiation. IL-6, gp80 and gp130 are upregulated in peripheral nerves, DRG and the spinal cord during experimental pain [99].

Numerous animal studies have implicated the potential involvement of IL-6 in neuropathic pain. IL-6 mRNA or protein is expressed at low or undetectable levels in normal sciatic nerves [100, 101] or DRG [102, 103], but expression is altered after injury. Increased levels of IL-6 mRNA [104], IL-6 receptors and IL-6R mRNA were detected following nerve injury [105]. It has been demonstrated that IL-6 is produced distal to the site of injury in sciatic nerve crush injury [101] by damaged neurons and glial cells [101, 102]. Intrathecal administration of human recombinant IL-6 has resulted in touch-evoked allodynia in normal rats [106]. Direct injection of IL-6 exerts hypersensitivity to thermal and mechanical stimuli [106–108] and IL-6 knockout mice have demonstrated diminished thermal sensitivity in a hot plate behavioural test [109, 110]. Sham-operated rats and rats without signs of allodynia following peripheral nerve injury have reduced numbers of IL-6-expressing cells in the sciatic nerve [111]. In addition, intrathecal administration of anti-IL-6 antibody attenuated mechanical allodynia in a rat model of peripheral injury [112]. In humans, increased levels of IL-6 mRNA have been reported in lumbar-herniated discs of patients with sciatic pain [113], and IL-6 was shown to be produced in a considerable amount following tissue injury and at the site of a surgical wound [114].

Interferon-γ

The IFN molecules are secreted soluble proteins of three classes: type I (α, β, ω and τ), type II (γ) and type III (λ). IFN-γ is the only type II IFN family member. It is a pro-inflammatory cytokine that has been implicated in the development of neuropathic pain in a number of animal models, and is likely to play a role in the production of many human pain states. Peripheral nerve injury models have demonstrated its release by astrocytes, damaged neurons and T cells within the dorsal horn of the spinal cord, an area explicitly involved in pain processing [115, 116]. Here the cytokine is believed to confer nociceptive hypersensitivity through the modulation of neuronal firing and the activation of other cell types, such as spinal microglia. IFN-γ contributes to central sensitisation through the induction of spontaneous firing of dorsal horn neurons, increasing wind-up of action potentials in response to repetitive electrical stimulation, as well as by reducing GABAergic inhibitory tone in the dorsal horn [117, 118]. The activation of spinal microglia is another mechanism by which IFN-γ promotes the development of neuropathic pain, and ablation of IFN-γ signalling impairs microglial activation and the development of pain hypersensitivity in mice with peripheral nerve injury [116, 119]. Stimulation of microglia with IFN-γ has been shown to induce an upregulation of proteins involved in the production of neuropathic pain. Some of these include Lyn tyrosine kinase, purinergic P2X4 receptor, iNOS and the chemokine receptor CCR2 [115, 119]. In support of the role of IFN-γ in the development of neuropathic pain, a study has shown that IFN-γ receptor knockout mice display reduced pain-related behaviours following peripheral nerve injury [120]. In addition, studies have further linked reduced neuropathic pain behaviours in peripheral nerve injury following treatments inducing IFN-γ downregulation with subsequent reductions in downstream modulators of pain, including the P2X4 receptor, iNOS and cyclooxygenase-2 [121, 122].

Interleukin-17

The IL-17 family consists of a subset of cytokines (IL-17A-F) that participate in host defence against microbial organisms and in the development of inflammatory diseases, with the most widely investigated cytokine of this family being IL-17A (also called IL-17). IL-17 is a pro-inflammatory cytokine produced by numerous cell types including T helper cells, cytotoxic T cells, neutrophils and glial cells [123–125]. Mice with peripheral nerve injury show increased levels of IL-17 in the spinal cord and injured sciatic nerve at time points corresponding with increased pain behaviours [126, 127]. A study demonstrated pain hypersensitivity 8 days following the induction of multiple models of neuropathic pain including CCI, partial sciatic nerve ligation and neuritis, and this was found to be associated with an upregulation of IL-17 in the injured sciatic nerve compared to intact sciatic nerves [128]. IL-17 knockout mice develop reduced pain hypersensitivity associated with decreased T cell and macrophage infiltration into injured sciatic nerves and the DRG, as well as decreased activation of microglia and astrocytes in the spinal cord [129]. These mice were subsequently shown to have decreased pro-inflammatory cell infiltration into the damaged sciatic nerve, as well as reduced expression of pro-inflammatory cytokines including TNF, IL-6 and IFN-γ [130]. Conversely, injection of IL-17 via intraplantar, intraneural (sciatic nerve), or intrathecal routes induced pain hypersensitivity in naïve mice [129].

Anti-Inflammatory Cytokines

Tumour Growth Factor-β

The TGF-β family in mammals includes three members, TGF-β1, 2 and 3, with the predominate isoform expressed in the immune system being TGF-β1. TGF-β1 has pleiotropic effects on adaptive immunity, especially in the regulation of effector and regulatory CD4+ T-cell responses. It has a strong anti-inflammatory effect and can mediate the development and function of immunosuppressive regulatory T cells. However, it can also enhance the development of IL-17-producing T cells that may be involved in autoimmune tissue injury [131].

Studies in animal models have provided strong evidence for the role of TGF-β1 in inhibiting neuropathic pain by suppressing neuroimmune responses of neurons and glia and promoting expression of endogenous opioids in the spinal cord [132]. Peripheral nerve injury causes downregulation of endogenous TGF-β1 in the lumbar spinal dorsal horn grey matter, ipsilateral to injury [133]. Treatment with TGF-β1 prevents the development of neuropathic pain following nerve injury and reverses previously established neuropathic pain [133, 134]. TGF-β1 suppresses peripheral nerve injury-induced neuropathic pain by inhibiting spinal microglia and astrocyte activation, neuronal damage and upregulation of pro-inflammatory cytokines such as TNF [133, 134]. Furthermore, TGF-β1 significantly inhibits nerve injury-induced downregulation of glutamate transporters, thus decreasing activity of spinal excitatory amino acids that may contribute to neuropathic pain [133]. In addition, intrathecal administration of flexibilide (derived from cultural soft coral) exerts an anti-inflammatory and analgesic effect through upregulation of spinal TGF-β in rats with neuropathic pain due to peripheral nerve injury [135].

Interleukin-10

IL-10, also known as human cytokine synthesis inhibitory factor, is an anti-inflammatory cytokine with pleiotropic effects on immunoregulation and inflammation. IL-10 downregulates type 1 T helper cell cytokine expression and is capable of inhibiting synthesis of pro-inflammatory cytokines. IL-10 activity has been shown to play a role in neuropathic pain. Early studies in animal models of peripheral nerve injury found that IL-10 attenuates neuropathic pain following peripheral nerve injury by suppressing macrophage recruitment and endoneurial activation of pro-inflammatory cytokines such as TNF [136]. These early findings have been confirmed by more recent animal studies examining the anti-inflammatory role of IL-10 in neuropathic pain. Micro-injections of IL-10 in the red nucleus suppressed mechanical allodynia in rats with peripheral nerve injury [137]. Herpes simplex virus-mediated expression of

IL-10 suppresses neuropathic pain induced by spinal cord injury [138] and HIV gp120 [139]. Repeated intrathecal delivery of IL-10 via plasmid DNA also abolished neuropathic pain after peripheral nerve injury [140]. The anti-nociceptive role of IL-10 in neuropathic pain has been further explored in therapeutic models, with gabapentin having been shown to enhance the anti-nociceptive effects of morphine in neuropathic pain by activating the IL-10-heme oxygenase-1 signalling pathway and inhibiting the expression of pro-inflammatory cytokines in the spinal cord following nerve injury [141].

Conclusion

In the previous section, we briefly reviewed the overwhelming evidence that demonstrates cytokine signalling plays a significant role in neuropathic pain. Neuropathic pain occurs as the result of a lesion or disease affecting the somatosensory nervous system, which always involves a component of neuroinflammation and cytokine signalling. Recent studies have highlighted the connection between cytokine signalling and neuropathic pain by providing evidence that several pro-inflammatory cytokines facilitate neuropathic pain, whereas other anti-inflammatory cytokines suppress neuropathic pain. Intriguingly, studies also point to alterations in the immune system, and more specifically cytokine signalling, in depression. The extent to which cytokine signalling contributes to depression associated with neuropathic pain is, however, less clear. In animals, simply injecting pro-inflammatory cytokines either systemically or locally within the central nervous system can induce both pain and depressive symptoms. This can partly be explained by the fact that pro-inflammatory cytokines precipitate the 'sickness response' that induces symptoms of both depression and hyperalgesia. Given the mounting evidence for their role in neuropathic pain and depression, cytokines might be used as aligned therapeutic targets. However, further investigations aimed at understanding cytokine signalling in pain responses and depressive symptoms may prove beneficial in designing novel therapeutic approaches to alleviate neuropathic pain and associated depression.

References

1 Treede RD, et al: Neuropathic pain: redefinition and a grading system for clinical and research purposes. Neurology 2008;70:1630–1635.
2 Costigan M, Scholz J, Woolf CJ: Neuropathic pain: a maladaptive response of the nervous system to damage. Annu Rev Neurosci 2009;32:1–32.
3 Campbell JN, Meyer RA: Mechanisms of neuropathic pain. Neuron 2006;52:77–92.
4 Jensen TS, et al: The clinical picture of neuropathic pain. Eur J Pharmacol 2001;429:1–11.
5 Woolf CJ, Mannion RJ: Neuropathic pain: aetiology, symptoms, mechanisms, and management. Lancet 1999;353:1959–1964.
6 Dworkin RH, et al: Advances in neuropathic pain: diagnosis, mechanisms, and treatment recommendations. Arch Neurol 2003;60:1524–1534.
7 Yawn BP, et al: The prevalence of neuropathic pain: clinical evaluation compared with screening tools in a community population. Pain Med 2009;10:586–593.

8 Torrance N, et al: The epidemiology of chronic pain of predominantly neuropathic origin. Results from a general population survey. J Pain 2006;7:281–289.
9 O'Connor AB: Neuropathic pain: quality-of-life impact, costs and cost effectiveness of therapy. Pharmacoeconomics 2009;27:95–112.
10 Meyer-Rosberg K, et al: Peripheral neuropathic pain – a multidimensional burden for patients. Eur J Pain 2001;5:379–389.
11 Finnerup NB, et al: Algorithm for neuropathic pain treatment: an evidence based proposal. Pain 2005; 118:289–305.
12 Attal N, et al: EFNS guidelines on the pharmacological treatment of neuropathic pain: 2010 revision. Eur J Neurol 2010;17:1113.e88.
13 Bennett GJ, Xie YK: A peripheral mononeuropathy in rat that produces disorders of pain sensation like those seen in man. Pain 1988;33:87–107.
14 Mogil JS, Davis KD, Derbyshire SW: The necessity of animal models in pain research. Pain 2010;151:12–17.
15 Mogil JS, Crager SE: What should we be measuring in behavioral studies of chronic pain in animals? Pain 2004;112:12–15.
16 Langford DJ, et al: Coding of facial expressions of pain in the laboratory mouse. Nat Methods 2010;7: 447–449.
17 Davoody L, et al: Conditioned place preference reveals tonic pain in an animal model of central pain. J Pain 2011;12:868–874.
18 Vallejo R, et al: The role of glia and the immune system in the development and maintenance of neuropathic pain. Pain Pract 2010;10:167–184.
19 Calvo M, Dawes JM, Bennett DLH: The role of the immune system in the generation of neuropathic pain. Lancet Neurol 2012;11:629–642.
20 Myers RR, Campana WM, Shubayev VI: The role of neuroinflammation in neuropathic pain: mechanisms and therapeutic targets. Drug Discov Today 2006;11:8–20.
21 Mueller M, et al: Macrophage response to peripheral nerve injury: the quantitative contribution of resident and hematogenous macrophages. Lab Invest 2003;83:175–185.
22 Perkins NM, Tracey DJ: Hyperalgesia due to nerve injury: role of neutrophils. Neuroscience 2000;101: 745–757.
23 Scholz J, Woolf CJ: The neuropathic pain triad: neurons, immune cells and glia. Nat Neurosci 2007;10: 1361–1368.
24 Shubayev VI, et al: TNFalpha-induced MMP-9 promotes macrophage recruitment into injured peripheral nerve. Mol Cell Neurosci 2006;31:407–415.
25 Martini R, et al: Interactions between Schwann cells and macrophages in injury and inherited demyelinating disease. Glia 2008;56:1566–1577.
26 Gaudet AD, Popovich PG, Ramer MS: Wallerian degeneration: gaining perspective on inflammatory events after peripheral nerve injury. J Neuroinflammation 2011;8:110.
27 Baron R, et al: Histamine-induced itch converts into pain in neuropathic hyperalgesia. Neuroreport 2001; 12:3475–3478.
28 Sahbaie P, et al: Role of substance P signaling in enhanced nociceptive sensitization and local cytokine production after incision. Pain 2009;145:341–349.
29 Zochodne DW, et al: Evidence for nitric oxide and nitric oxide synthase activity in proximal stumps of transected peripheral nerves. Neuroscience 1999;91: 1515–1527.
30 Moalem G, Tracey DJ: Immune and inflammatory mechanisms in neuropathic pain. Brain Res Rev 2006;51:240–264.
31 Miller RJ, et al: Cytokine and chemokine regulation of sensory neuron function. Handb Exp Pharmacol 2009, pp 417–449.
32 Moalem G, Xu K, Yu L: T lymphocytes play a role in neuropathic pain following peripheral nerve injury in rats. Neuroscience 2004;129:767–777.
33 Costigan M, et al: T-cell infiltration and signaling in the adult dorsal spinal cord is a major contributor to neuropathic pain-like hypersensitivity. J Neurosci 2009;29:14415–14422.
34 Kleinschnitz C, et al: T cell infiltration after chronic constriction injury of mouse sciatic nerve is associated with interleukin-17 expression. Exp Neurol 2006;200:480–485.
35 Milligan ED, Watkins LR: Pathological and protective roles of glia in chronic pain. Nat Rev Neurosci 2009;10:23–36.
36 Leo RJ: Chronic pain and comorbid depression. Curr Treat Options Neurol 2005;7:403–412.
37 Gustorff B, et al: Prevalence of self-reported neuropathic pain and impact on quality of life: a prospective representative survey. Acta Anaesthesiol Scand 2008;52:132–136.
38 Holzberg AD, et al: The effects of depression and chronic pain on psychosocial and physical functioning. Clin J Pain 1996;12:118–125.
39 Fuss I, et al: Prognostic factors for pain relief and functional improvement in chronic pain after inpatient rehabilitation. Clin J Pain 2014;30:279–285.
40 McWilliams LA, Cox BJ, Enns MW: Mood and anxiety disorders associated with chronic pain: an examination in a nationally representative sample. Pain 2003;106:127–133.
41 Boulanger L, et al: Impact of comorbid depression or anxiety on patterns of treatment and economic outcomes among patients with diabetic peripheral neuropathic pain. Curr Med Res Opin 2009;25:1763–1773.

42 Volpi A, et al: Clinical and psychosocial correlates of post-herpetic neuralgia. J Med Virol 2008;80:1646–1652.
43 Hart BL: Biological basis of the behavior of sick animals. Neurosci Biobehav Rev 1988;12:123–137.
44 Maier SF, et al: Interleukin-1 mediates the behavioral hyperalgesia produced by lithium chloride and endotoxin. Brain Res 1993;623:321–324.
45 Watkins LR, Maier SF: Immune regulation of central nervous system functions: from sickness responses to pathological pain. J Intern Med 2005;257:139–155.
46 Dowlati Y, et al: A meta-analysis of cytokines in major depression. Biol Psychiatry 2010;67:446–457.
47 Howren MB, Lamkin DM, Suls J: Associations of depression with C-reactive protein, IL-1, and IL-6: a meta-analysis. Psychosom Med 2009;71:171–186.
48 Maletic V, Raison CL: Neurobiology of depression, fibromyalgia and neuropathic pain. Front Biosci (Landmark Ed) 2009;14:5291–5338.
49 Plotsky PM, Owens MJ, Nemeroff CB: Psychoneuroendocrinology of depression. Hypothalamic-pituitary-adrenal axis. Psychiatr Clin North Am 1998;21: 293–307.
50 Ressler KJ, Nemeroff CB: Role of serotonergic and noradrenergic systems in the pathophysiology of depression and anxiety disorders. Depress Anxiety 2000;12(suppl 1):2–19.
51 Berton O, Nestler EJ: New approaches to antidepressant drug discovery: beyond monoamines. Nat Rev Neurosci 2006;7:137–151.
52 Benedetti M, et al: Plasma corticosterone levels in mouse models of pain. Eur J Pain 2012;16:803–815.
53 Raison CL, Capuron L, Miller AH: Cytokines sing the blues: inflammation and the pathogenesis of depression. Trends Immunol 2006;27:24–31.
54 Bonaccorso S, et al: Increased depressive ratings in patients with hepatitis C receiving interferon-alpha-based immunotherapy are related to interferon-alpha-induced changes in the serotonergic system. J Clin Psychopharmacol 2002;22:86–90.
55 Capuron L, et al: Neurobehavioral effects of interferon-alpha in cancer patients: phenomenology and paroxetine responsiveness of symptom dimensions. Neuropsychopharmacology 2002;26:643–652.
56 Dunn AJ, Swiergiel AH, de Beaurepaire R: Cytokines as mediators of depression: what can we learn from animal studies? Neurosci Biobehav Rev 2005;29: 891–909.
57 Maes M, et al: The new '5-HT' hypothesis of depression: cell-mediated immune activation induces indoleamine 2,3-dioxygenase, which leads to lower plasma tryptophan and an increased synthesis of detrimental tryptophan catabolites (TRYCATs), both of which contribute to the onset of depression. Prog Neuropsychopharmacol Biol Psychiatry 2011; 35:702–721.
58 Miller AH, Maletic V, Raison CL: Inflammation and its discontents: the role of cytokines in the pathophysiology of major depression. Biol Psychiatry 2009;65:732–741.
59 John CD, Buckingham JC: Cytokines: regulation of the hypothalamo-pituitary-adrenocortical axis. Curr Opin Pharmacol 2003;3:78–84.
60 Locksley RM, et al: The TNF and TNF receptor superfamilies: integrating mammalian biology. Cell 2001;104:487–501.
61 Thalayasingam N, Isaacs JD: Anti-TNF therapy. Best Pract Res Clin Rheumatol 2011;25:549–567.
62 Horiuchi T, et al: Transmembrane TNF-alpha: structure, function and interaction with anti-TNF agents. Rheumatology (Oxford, England) 2010;49: 1215–1228.
63 Macewan DJ: TNF receptor subtype signalling: differences and cellular consequences. Cell Signal 2002; 14:477–492.
64 Empl M, et al: TNF-alpha expression in painful and nonpainful neuropathies. Neurology 2001;56:1371–1377.
65 Uçeyler N, et al: Differential expression of cytokines in painful and painless neuropathies. Neurology 2007;69:42–49.
66 George A, Buehl A, Sommer C: Wallerian degeneration after crush injury of rat sciatic nerve increases endo- and epineurial tumor necrosis factor-alpha protein. Neurosci Lett 2004;372:215–219.
67 Wagner R, Myers R: Endoneurial injection of TNF-α produces neuropathic pain behaviors. Neuroreport 1996;7:2897–2901.
68 Sorkin LS, Doom CM: Epineurial application of TNF elicits an acute mechanical hyperalgesia in the awake rat. J Peripher Nerv Syst 2000;100:96–100.
69 Sommer C, et al: Etanercept reduces hyperalgesia in experimental painful neuropathy. J Peripher Nerv Syst 2001;6:67–72.
70 Sommer C, Marziniak M, Myers RR: The effect of thalidomide treatment on vascular pathology and hyperalgesia caused by chronic constriction injury of rat nerve. Pain 1998;74:83–91.
71 Nadeau S, et al: Functional recovery after peripheral nerve injury is dependent on the pro-inflammatory cytokines IL-1β and TNF: implications for neuropathic pain. J Neurosci 2011;31:12533–12542.
72 Gao Y-J, Zhang L, Ji R-R: Spinal injection of TNF-α-activated astrocytes produces persistent pain symptom mechanical allodynia by releasing monocyte chemoattractant protein-1. Glia 2011;58:1871–1880.
73 Ohtori S, et al: TNF-α and TNF-α receptor type 1 upregulation in glia and neurons after peripheral nerve injury studies in murine DRG and spinal cord. Spine 2004;29:1082–1088.
74 Xu J-T, et al: The role of tumor necrosis factor-alpha in the neuropathic pain induced by lumbar 5 ventral root transection in rat. Pain 2006;123:306–321.

75 Zhang H, Zhang H, Dougherty PM: Dynamic effects of TNF-α on synaptic transmission in mice over time following sciatic nerve chronic constriction injury. J Neurophysiol 2013;110:1663–1671.
76 Leung L, Cahill C: TNF-α and neuropathic pain – a review. J Neuroinflammation 2010;7:27.
77 Dellarole A, et al: Neuropathic pain-induced depressive-like behavior and hippocampal neurogenesis and plasticity are dependent on TNFR1 signaling. Brain Behav Immun 2014;41:65–81.
78 Dinarello CA: Biologic basis for interleukin-1 in disease. Blood 1996;87:2095–2147.
79 Rothwell NJ, Luheshi GN: Interleukin 1 in the brain: biology, pathology and therapeutic target. Trends Neurosci 2000;23:618–625.
80 Yu B, Shinnick-Gallagher P: Interleukin-1 beta inhibits synaptic transmission and induces membrane hyperpolarization in amygdala neurons. J Pharmacol Exp Ther 1994;271:590–600.
81 Samad TA, et al: Interleukin-1beta-mediated induction of Cox-2 in the CNS contributes to inflammatory pain hypersensitivity. Nature 2001;410:471–475.
82 DeLeo JA, Colburn RW, Rickman AJ: Cytokine and growth factor immunohistochemical spinal profiles in two animal models of mononeuropathy. Brain Res 1997;759:50–57.
83 Sweitzer SM, et al: Acute peripheral inflammation induces moderate glial activation and spinal IL-1beta expression that correlates with pain behavior in the rat. Brain Res 1999;829:209–221.
84 Ferreira SH, et al: Interleukin-1 beta as a potent hyperalgesic agent antagonized by a tripeptide analogue. Nature 1988;334:698–700.
85 Watkins LR, et al: Characterization of cytokine-induced hyperalgesia. Brain Res 1994;654:5–26.
86 Oka T, Aou S, Hori T: Intracerebroventricular injection of interleukin-1 beta induces hyperalgesia in rats. Brain Res 1993;624:61–68.
87 Reeve AJ, et al: Intrathecally administered endotoxin or cytokines produce allodynia, hyperalgesia and changes in spinal cord neuronal responses to nociceptive stimuli in the rat. Eur J Pain 2000;4:247–257.
88 Sung CS, et al: Intrathecal interleukin-1beta administration induces thermal hyperalgesia by activating inducible nitric oxide synthase expression in the rat spinal cord. Brain Res 2004;1015:145–153.
89 Wu J, et al: Nitric oxide synthase in spinal cord central sensitization following intradermal injection of capsaicin. Pain 2001;94:47–58.
90 Sung CS, et al: Inhibition of p38 mitogen-activated protein kinase attenuates interleukin-1beta-induced thermal hyperalgesia and inducible nitric oxide synthase expression in the spinal cord. J Neurochem 2005;94:742–752.
91 Tsuda M, et al: Activation of p38 mitogen-activated protein kinase in spinal hyperactive microglia contributes to pain hypersensitivity following peripheral nerve injury. Glia 2004;45:89–95.
92 Chessell IP, et al: Disruption of the P2X7 purinoceptor gene abolishes chronic inflammatory and neuropathic pain. Pain 2005;114:386–396.
93 Binshtok AM, et al: Nociceptors are interleukin-1beta sensors. J Neurosci 2008;28:14062–14073.
94 Wolf G, et al: Genetic impairment of interleukin-1 signaling attenuates neuropathic pain, autotomy, and spontaneous ectopic neuronal activity, following nerve injury in mice. Pain 2006;120:315–324.
95 Gabay E, et al: Chronic blockade of interleukin-1 (IL-1) prevents and attenuates neuropathic pain behavior and spontaneous ectopic neuronal activity following nerve injury. Eur J Pain 2011;15:242–248.
96 Hirota H, et al: Accelerated nerve regeneration in mice by upregulated expression of interleukin (IL) 6 and IL-6 receptor after trauma. J Exp Med 1996; 183:2627–2634.
97 Heinrich PC, et al: Interleukin-6-type cytokine signalling through the gp130/Jak/STAT pathway. Biochem J 1998;334:297–314.
98 Rose-John S, et al: Interleukin-6 biology is coordinated by membrane-bound and soluble receptors: role in inflammation and cancer. J Leukoc Biol 2006;80:227–236.
99 De Jongh RF, et al: The role of interleukin-6 in nociception and pain. Anesth Analg 2003;96:1096–1103.
100 Kurek JB, et al: Up-regulation of leukaemia inhibitory factor and interleukin-6 in transected sciatic nerve and muscle following denervation. Neuromuscul Disord 1996;6:105–114.
101 Bolin LM, et al: Interleukin-6 production by Schwann cells and induction in sciatic nerve injury. J Neurochem 1995;64:850–858.
102 Murphy PG, et al: Induction of interleukin-6 in axotomized sensory neurons. J Neurosci 1995;15: 5130–5138.
103 Nordlind K, et al: Expression of interleukin-6 in human dorsal root ganglion cells. Neurosci Lett 2000;280:139–142.
104 Martucci C, et al: The purinergic antagonist PPADS reduces pain related behaviours and interleukin-1 beta, interleukin-6, iNOS and nNOS overproduction in central and peripheral nervous system after peripheral neuropathy in mice. Pain 2008;137:81–95.
105 Ito Y, et al: Differential temporal expression of mRNAs for ciliary neurotrophic factor (CNTF), leukemia inhibitory factor (LIF), interleukin-6 (IL-6), and their receptors (CNTFR alpha, LIFR beta, IL-6R alpha and gp130) in injured peripheral nerves. Brain Res 1998;793:321–327.

106 DeLeo JA, et al: Interleukin-6-mediated hyperalgesia/allodynia and increased spinal IL-6 expression in a rat mononeuropathy model. J Interferon Cytokine Res 1996;16:695–700.
107 Oka T, et al: Intracerebroventricular injection of interleukin-6 induces thermal hyperalgesia in rats. Brain Res 1995;692:123–128.
108 Brenn D, Richter F, Schaible HG: Sensitization of unmyelinated sensory fibers of the joint nerve to mechanical stimuli by interleukin-6 in the rat: an inflammatory mechanism of joint pain. Arthritis Rheum 2007;56:351–359.
109 Ramer MS, et al: Spinal nerve lesion-induced mechanoallodynia and adrenergic sprouting in sensory ganglia are attenuated in interleukin-6 knockout mice. Pain 1998;78:115–121.
110 Zhong J, et al: Sensory impairments and delayed regeneration of sensory axons in interleukin-6-deficient mice. J Neurosci 1999;19:4305–4313.
111 Cui JG, et al: Possible role of inflammatory mediators in tactile hypersensitivity in rat models of mononeuropathy. Pain 2000;88:239–248.
112 Arruda JL, et al: Intrathecal anti-IL-6 antibody and IgG attenuates peripheral nerve injury-induced mechanical allodynia in the rat: possible immune modulation in neuropathic pain. Brain Res 2000; 879:216–225.
113 Andrade P, et al: Elevated IL-1beta and IL-6 levels in lumbar herniated discs in patients with sciatic pain. Eur Spine J 2013;22:714–720.
114 Holzheimer RG, Steinmetz W: Local and systemic concentrations of pro- and anti-inflammatory cytokines in human wounds. Eur J Med Res 2000;5: 347–355.
115 Racz I, et al: Interferon-gamma is a critical modulator of CB(2) cannabinoid receptor signaling during neuropathic pain. J Neurosci 2008;28:12136–12145.
116 Costigan M, et al: T-cell infiltration and signaling in the adult dorsal spinal cord is a major contributor to neuropathic pain-like hypersensitivity. J Neurosci 2009;29:14415–14422.
117 Vikman KS, Siddall PJ, Duggan AW: Increased responsiveness of rat dorsal horn neurons in vivo following prolonged intrathecal exposure to interferon-gamma. Neuroscience 2005;135:969–977.
118 Vikman KS, Duggan AW, Siddall PJ: Interferon-gamma induced disruption of GABAergic inhibition in the spinal dorsal horn in vivo. Pain 2007; 133:18–28.
119 Tsuda M, et al: IFN-gamma receptor signaling mediates spinal microglia activation driving neuropathic pain. Proc Natl Acad Sci U S A 2009;106: 8032–8037.
120 Robertson B, et al: Interferon-gamma receptors in nociceptive pathways: role in neuropathic pain-related behaviour. Neuroreport 1997;8:1311–1316.
121 Chen XM, et al: Electroacupuncture inhibits excessive interferon-gamma evoked up-regulation of P2X4 receptor in spinal microglia in a CCI rat model for neuropathic pain. Br J Anaesth 2015;114:150–157.
122 Jean YH, et al: Capnellene, a natural marine compound derived from soft coral, attenuates chronic constriction injury-induced neuropathic pain in rats. Br J Pharmacol 2009;158:713–725.
123 Kawanokuchi J, et al: Production and functions of IL-17 in microglia. J Neuroimmunol 2008;194:54–61.
124 Tzartos JS, et al: Interleukin-17 production in central nervous system-infiltrating T cells and glial cells is associated with active disease in multiple sclerosis. Am J Pathol 2008;172:146–155.
125 Ferretti S, et al: IL-17, produced by lymphocytes and neutrophils, is necessary for lipopolysaccharide-induced airway neutrophilia: IL-15 as a possible trigger. J Immunol 2003;170:2106–2112.
126 Li J, et al: Nerve injury-related autoimmunity activation leads to chronic inflammation and chronic neuropathic pain. Anesthesiology 2013;118:416–429.
127 Kleinschnitz C, et al: T cell infiltration after chronic constriction injury of mouse sciatic nerve is associated with interleukin-17 expression. Exp Neurol 2006;200:480–485.
128 Noma N, et al: Interleukin-17 levels in rat models of nerve damage and neuropathic pain. Neurosci Lett 2011;493:86–91.
129 Kim CF, Moalem-Taylor G: Interleukin-17 contributes to neuroinflammation and neuropathic pain following peripheral nerve injury in mice. J Pain 2011;12:370–383.
130 Day YJ, et al: Lack of interleukin-17 leads to a modulated micro-environment and amelioration of mechanical hypersensitivity after peripheral nerve injury in mice. Pain 2014;155:1293–1302.
131 Bettelli E, et al: Reciprocal developmental pathways for the generation of pathogenic effector TH17 and regulatory T cells. Nature 2006;441:235–238.
132 Lantero A, et al: Transforming growth factor-beta in normal nociceptive processing and pathological pain models. Mol Neurobiol 2012;45:76–86.
133 Chen NF, et al: TGF-beta1 attenuates spinal neuroinflammation and the excitatory amino acid system in rats with neuropathic pain. J Pain 2013;14:1671–1685.
134 Echeverry S, et al: Transforming growth factor-beta1 impairs neuropathic pain through pleiotropic effects. Mol Pain 2009;5:16.
135 Chen NF, et al: Flexibilide obtained from cultured soft coral has anti-neuroinflammatory and analgesic effects through the upregulation of spinal transforming growth factor-beta1 in neuropathic rats. Mar Drugs 2014;12:3792–3817.

136 Wagner R, Janjigian M, Myers RR: Anti-inflammatory interleukin-10 therapy in CCI neuropathy decreases thermal hyperalgesia, macrophage recruitment, and endoneurial TNF-alpha expression. Pain 1998;74:35–42.
137 Wang ZH, et al: Interleukin-10 of red nucleus plays anti-allodynia effect in neuropathic pain rats with spared nerve injury. Neurochem Res 2012;37:1811–1819.
138 Lau D, et al: Herpes simplex virus vector-mediated expression of interleukin-10 reduces below-level central neuropathic pain after spinal cord injury. Neurorehabil Neural Repair 2012;26:889–897.
139 Zheng W, et al: Interleukin 10 mediated by herpes simplex virus vectors suppresses neuropathic pain induced by human immunodeficiency virus gp120 in rats. Anesth Analg 2014;119:693–701.
140 Milligan ED, et al: Repeated intrathecal injections of plasmid DNA encoding interleukin-10 produce prolonged reversal of neuropathic pain. Pain 2006;126:294–308.
141 Bao YH, et al: Gabapentin enhances the morphine anti-nociceptive effect in neuropathic pain via the interleukin-10-heme oxygenase-1 signalling pathway in rats. J Mol Neurosci 2014;54:137–146.

Dr. Gila Moalem-Taylor
School of Medical Sciences, The University of New South Wales
Wallace Wurth Building East, Level 3, Room 327
Sydney, NSW 2052 (Australia)
E-Mail gila@unsw.edu.au

Finn DP, Leonard BE (eds): Pain in Psychiatric Disorders. Mod Trends Pharmacopsychiatry.
Basel, Karger, 2015, vol 30, pp 67–79 (DOI: 10.1159/000435933)

Biogenic Amines and the Amino Acids GABA and Glutamate: Relationships with Pain and Depression

Curtis Benson[a, b] • Katherine Mifflin[a, b] • Bradley Kerr[a–c] •
Sam J.B. Jesudasan[b, c] • Serdar Dursun[b, c] • Glen Baker[b, c]

[a]Department of Anesthesiology and Pain Medicine, [b]Neuroscience and Mental Health Institute, and
[c]Neurochemical Research Unit, Department of Psychiatry, University of Alberta, Edmonton, Alta., Canada

Abstract

Although it is well known that there is a high degree of comorbidity between chronic pain and mood and anxiety disorders, the mechanisms involved in these co-occurrences are not clear. It appears that numerous neurotransmitters and neuromodulators are involved, and this chapter focuses on the monoamine neurotransmitters noradrenaline, 5-hydroxytryptamine (5-HT, serotonin), and dopamine and the amino acid neurotransmitters GABA (γ-aminobutyric acid) and glutamate in chronic pain and depression. Numerous preclinical and clinical neurochemical, neuroanatomical, pharmacological and molecular biological studies as well as clinical pharmacological treatment investigations implicate noradrenaline, 5-HT and, to a lesser extent, dopamine in the etiology of pain and depression. Similarly, preclinical and clinical studies on GABAergic and glutamatergic mechanisms as well as reports on the actions of neuroactive steroids suggest that GABA and glutamate play an important role in the etiology of pain and depression and may contribute to comorbidity.

Chronic pain is often debilitating, and lifetime prevalences have been reported to range from 20 to 60% [1, 2]. There can be a large impact on the everyday life of people suffering from chronic pain as seen in a reduction in quality of life, disrupted mood, altered sleep and cognitive dysfunction [3–6]. There is often a high degree of comorbidity between chronic pain and psychiatric disorders, including depression and anxiety [7–12]. The co-existence of depression and chronic pain often results in increased duration and severity of depression and its physical symptoms, a greater risk of relapse, a lack of patient satisfaction with treatment, and a poor treatment response [7, 10, 13, 14]. The relationship between depression and chronic pain is so well established that it has been named the 'depression-pain syndrome' or the 'pain/depression dyad' [15, 16]. Chronic pain and depression co-occur, but also promote the development of each other, i.e. chronic pain can be a strong predictor of subsequently developing major depressive disorder and vice versa [17]. Despite the frequency of this co-occurrence of pain and depression, the reasons for it are not entirely clear. Yalcin et al. [12] have reviewed the literature on preclinical modelling of the anxiodepressive consequences of neuropathic pain and have discussed some of the problems with investigating such comorbidity, including the variety of neurotransmitters and neuromodulators that are probably involved, time points at which experiments were conducted, species and strains of animals used, pain models used, and the time of the day-night cycle when the animals are tested.

Often patients with chronic pain are treated with antidepressants [18–24]. Most of these drugs also have analgesic effects in people who are not depressed, and the pain-relieving effects appear to be independent of the antidepressant effects [25, 26]. Most of the antidepressants have actions that result in increased functional availability of the biogenic amines noradrenaline (NA) and/or 5-hydroxytryptamine (5-HT, serotonin) [27–30], and evidence has now accumulated indicating that GABAergic and glutamatergic mechanisms are also involved in the actions of these drugs [31–34]. These four neurotransmitter systems have also been implicated in the development and/or regulation of pain [12, 35–42]. This chapter will focus on the possible role of the biogenic amine neurotransmitters NA and serotonin (and to a lesser extent dopamine) and the neurotransmitter amino acids GABA (γ-aminobutyric acid) and glutamate in pain and depression.

The Role of Monoamines in Depression and Chronic Pain

Preclinical and clinical research in chronic pain has shown that altered spinal serotonergic and noradrenergic transmission contribute to pain in depression [43, 44]. Serotoninergic and noradrenergic descending neurones project from the brain stem to the spinal cord's dorsal horn where a complex set of biochemical actions can take place involving other neurotransmitters/neuromodulators such as GABA, glutamate and substance P [26]. Because of the multiple 5-HT receptors which can modulate nociceptive transmission, the descending 5-HT pathways may exert either inhibitory or excitatory effects on pain perception, while noradrenergic neurones projecting to the dorsal horn produce mostly antinociceptive actions (see Matsuzawa-Yanagida et al. [24] for a summary of the effects of 5-HT and NA in the dorsal horn and the receptors involved).

Serotonergic and noradrenergic systems both also send ascending projection fibers to supraspinal areas involved in control of emotional response (possibly affecting depression) [45, 46]. Imaging studies have shown that brain areas involved in integration of the emotional and sensory aspects of pain can be modulated by depression [47]. For example, Bär et al. [48] have shown that there is hyperactivation of the ventrolateral and dorsolateral prefrontal cortices, which may contribute to thermal hypoalgesia, a sensory dysregulation phenomenon associated with depression. Also, activation in the amygdala and the anterior insular cortex are known to be different in fibromyalgia patients with and without major depression, suggesting that these areas are altered in comorbid depression and pain patients [49]. The anterior cingulate cortex, an integration center interconnecting neurones from the frontal cortex, thalamus and amygdala and involved in processing cognitive, emotional and autonomic functions [50, 51], is implicated in the pathophysiology of both depression and pain processing [12]. A comprehensive overview of brain areas, including the cortex, hippocampus, amygdala and mesolimibic system, that may be involved in the anxiodepressive consequences of neuropathic pain is provided in the review article by Yalcin et al. [12].

Similar changes in tyrosine hydroxylase, NA transport and α_2-adrenoreceptor levels in noradrenergic neurones in the locus ceoruleus (LC) have been observed in animal models of neuropathic pain and depression, supporting a possible key role of the LC in the relationship between pain and depression [52–55]. Interestingly, although it is known that NA blocks nociceptive transmission in the spinal cord, both antinociception and pronociception have been reported in the brain [56].

5-HT's effect on pain processing is complex [57, 58]. Initially characterized as a key neurotransmitter (along with NA) in brainstem structures that send descending inhibitory inputs to the spinal cord, 5-HT is now also recognized for its capacity to facilitate pain transmission [59, 60]. These discrepant excitatory/inhibitory responses to 5-HT are mediated through activation of distinct receptor subtypes for 5-HT where certain classes mediate predominantly excitatory actions (e.g. 5-HT_2, 5-HT_3) while others mediate its inhibitory effects (e.g. 5-HT_1) [61]. The 5-HT_1 receptor agonist DPAT has been reported to produce pronociceptive effects through restriction of GABA release [62], the 5-HT_3 receptor antagonist alosteron reverses mechanical somatic hypersensitivity and prevents development of visceral hyperalgesia [63], and the 5-HT_4 receptor agonist tegaserod produces analgesia by triggering release of opioids at supraspinal sites, thus activating the descending noradrenergic pathways to the spinal cord [64].

It is noteworthy that many of the antidepressants used for the treatment of neuropathic pain [tricyclic antidepressants (TCAs), e.g. amitriptyline, serotonin-noradrenaline reuptake inhibitors (SNRIs), e.g. duloxetine, and selective serotonin reuptake inhibitors (SSRIs), e.g. paroxetine] act on the 5-HT system, blocking reuptake mechanisms to elevate levels of 5-HT in the synaptic cleft [65]. However, it is important to note that the first two classes of antidepressants (TCAs and SNRIs), which also inhibit reuptake of NA, are often more effective than SSRIs for treating pain [12, 26].

Some authors have indicated that monoamine oxidase inhibitors (MAOIs) should not be used in treating pain disorders [66, 67], but the MAOI antidepressant phenelzine (PLZ) has been reported to be effective in treating pain associated with depression [68, 69]. This drug causes a dramatic increase in brain and spinal cord levels of 5-HT and NA [70, 71], and also appears to reduce pain in the experimental autoimmune encephalomyelitis animal model of multiple sclerosis [Benson, Mifflin, Baker and Kerr, unpublished], a disorder which is often accompanied by pain. However, this drug also inhibits GABA transaminase and elevates brain and spinal cord levels of GABA [71, 72], which may contribute to anxiolytic, antidepressant and/or pain-attenuating effects. Interestingly, PLZ and its metabolite, phenylethylidenehydrazine (PEH), also interact with toxic aldehydes like acrolein, which are known to contribute to inflammation and pain [73, 74], to reduce their activity by converting them to hydrazones [75].

Although the main focus of research on the biogenic amines and their involvement in pain and depression has been on NA and 5-HT, there is a growing body of evidence in support of a role for dopamine in perception and modulation of pain [76–78]. Hypersensitivity and hyposensitivity to pain occurs respectively in subjects with Parkinson's disease (low dopamine levels in the brain) and schizophrenia (hypersensitive mesolimbic dopaminergic system) [77–79]. Studies in animals and humans have suggested the importance of dopamine D_2 receptors in modulation of pain sensation in the striatum [80]. The TCAs have very little effect on dopamine reuptake, but it has been proposed that their adrenergic effect may result in an indirect dopaminergic effect through desensitization of dopamine D_2 receptors [18]. The atypical antidepressant bupropion (inhibits reuptake of NA and dopamine) has been used to treat neuropathic pain [81].

GABA and Glutamate in Depression and Chronic Pain

GABA and glutamate are major inhibitory and excitatory neurotransmitters, respectively, and it is well known that the delicate balance between them is disrupted in a number of psychiatric and neurological disorders [82–86]. Hypofunction of the GABAergic system [33, 87] and hyperfunction of the glutamatergic system [31, 32, 34] have been reported in depression.

A loss of effective inhibitory neurotransmission in the spinal dorsal horn is an important mechanism leading to the development of neuropathic pain. Peripheral nerve injury models associated with neuropathic pain behaviors are associated with significant reductions in GABA-mediated inhibitory postsynaptic potentials in dorsal horn neurons [38]. A selective loss of GABAergic interneurons through apoptosis after peripheral nerve injury has been proposed as the underlying cause for the lost inhibitory signaling, although the precise mechanism remains a matter of debate [88, 89]. Irrespective of its underlying cause, reduced inhibitory neurotransmission is recognized as a critical feature of chronic pain states. Both GABA-A and GABA-B receptors appear to be involved in the pathophysiology of chronic pain [90]. There is now considerable interest in using positive allosteric modulators of both types of GABA receptors as possible treatments for various types of pain, including visceral pain since they have been shown to be effective in animal models (reviewed in Kannampalli and Sengupta [91]) and to have an improved side-effect profile compared to direct agonists of these receptors. The TCA imipramine and the MAOI phenelzine, both antidepressants that have been used for treating pain, have been reported to alter GABA-A receptor subunit gene expression in the rat brainstem [92].

The gabapentinoid anticonvulsants gabapentin and pregabalin are used frequently to treat some pain syndromes [93, 94]. They are derivatives of GABA, but seem to have very little direct effect on GABA-A or GABA-B receptors [95] and their mechanism of action is still a matter of debate [96]. However, they bind to $\alpha_2\delta$-auxiliary subunits of voltage-dependent calcium channels [97]. Using a murine chronic pain model, Tanabe et al. [96] reported that gabapentin and pregabalin produced a supraspinally mediated analgesic effect that involved central and spinal NA; they also proposed that gabapentin presynaptically reduces GABAergic transmission, thus removing the inhibitory influence on the LC in neuropathic pain states and resulting in activation of the descending noradrenergic system. Hayashida et al. [98] suggested that gabapentin acts directly on the LC through a glutamate-dependent mechanism to stimulate descending noradrenergic inhibition to reduce hypersensitivity after peripheral nerve injury; they also reported that intravenous administration of gabapentin to rats induced NA release in the spinal dorsal horn. Yoshizumi et al. [99] reported that gabapentin inhibits GABA release in the LC, but not in the spinal dorsal horn following peripheral nerve injury in rats. In a study using in vivo microdialysis in the LC of spinal nerve-ligated rats, Suto et al. [100] obtained results which suggested that gabapentin induced glutamate release from astrocytes in the LC via glutamate transporter-1-dependent mechanisms, resulting in stimulation of descending inhibition; they also indicated that this target of gabapentin in astrocytes does not have a requirement for interaction with $\alpha_2\delta$-subunits in neurons.

It is well established in a number of animal models of chronic pain that the development of allodynia and/or hyperalgesia is critically dependent on enhanced signaling at excitatory glutamatergic receptor complexes [101–104]. NMDA receptors, AMPA receptors and some metabotropic glutamate receptors may be involved in modulation of pain, and this topic is reviewed comprehensively in articles by Bleakman et al. [103] and Kannampalli and Sengupta [91]. Recent research into the involvement of glutamate in chronic pain and depression has shown that glutamate receptors in the nucleus accumbens (an area of the brain affected in both pain and depression) may contribute to the development of depressive-like behaviors associated with chronic pain [105]. The study found that in a rat model of spared nerve injury (a common model of chronic pain), rats displayed comorbid depression and pain as demonstrated by a decreased preference for sucrose and increased immobility in the forced swim test [105]. Interestingly, it was found that there was an increase in the synapse of GluA1 AMPA receptor subunits in the nucleus accumbens of these rats, which led to the formation of calcium permeable AMPA receptors [105]. This is important as normally there are fewer of these calcium permeable receptors compared to nonpermeable AMPA receptors in this area. Surprisingly, this was found to be a compensatory mechanism for depression associated with chronic pain. It was determined that the blockade of these receptors worsened depressive symptoms, while their potentiation ameliorated depressive symptoms [105]. Thus, it appears as though the increase in calcium-permeable AMPA receptors has a protective role in the development of depression related to chronic pain. Interestingly, although changes in the level of these receptors improved depressive symptoms, the associated pain was not reduced. This suggests that there may be some endogenous compensatory mechanisms to help manage depression associated with chronic pain. This study revealed a new potential therapeutic target for the treatment of depression associated with chronic pain through the increased amplification of calcium-permeable AMPA receptor activity. Interestingly, the NMDA receptor antagonist and dissociative anesthetic ketamine, when infused at subanesthetic doses, is a rapid acting antidepressant [106, 107] and has also been used for treating various therapy-resistant chronic pain syndromes [108].

The possible importance of GABA and glutamate in pain and depression is also supported by studies done on neuroactive steroids (reviewed in detail in the chapter by Mifflin et al. [this vol., pp. 94–102]), rapid acting steroids that act on a number of receptors, particularly GABA-A and NMDA receptors [109–113]. It has been reported that several of these neuroactive steroids [e.g. allopregnanolone, progesterone and THDOC (positive allosteric modulators at the GABA-A receptor)] have analgesic properties in various pain syndromes [114–119]. Several antidepressants increase brain and body fluid levels of allopregnanolone [120–124], and allopregnanolone has been reported to have antidepressant and anxiolytic properties as well as to produce analgesic effects [125].

Antidepressants in the Treatment of Chronic Pain

Given that monoaminergic systems are involved in both chronic pain and depression, it is not surprising that commonly used treatments for chronic pain are antidepressants. However, it should also be remembered that most of these antidepressants are also effective at attenuating pain in people without depression [25, 26], i.e. they have a true analgesic effect. Antidepressants are currently used for the treatment of pain conditions such as arthritis, central pain syndrome, fibromyalgia, low back pain, migraine, diabetic neuropathy and postherpetic neuralgia (commonly known as shingles-associated pain) in addition to their typical use in mood disorders [22, 67]. These pain conditions are most commonly treated with TCAs which work by inhibiting the reuptake of 5-HT and NA. Commonly used TCAs in chronic pain are amitriptyline, imipramine, desipramine, doxepin and nortriptyline [67].

Although TCAs are well established for use in the treatment of chronic pain, they were replaced many years ago by SSRIs and SNRIs as first-line drugs for treating depression. SSRIs are arguably the most commonly used class of antidepressants for the treatment of depression, yet they appear to be inferior to TCAs and SNRIs (both of which inhibit NA reuptake) with regard to analgesic effects [67, 126]. Thus, although there are certainly underlying common biological mechanisms with regard to monoaminergic systems between chronic pain and depression, there do appear to be some differences as well. The relative lack of responsiveness to SSRIs and relative success of TCAs in chronic pain patients suggests that noradrenergic pathways may be more important in chronic pain compared to depression. This idea is further supported by evidence showing that SNRIs are also more effective than SSRIs in the management of chronic pain [127–132]. However, it should also be remembered that TCAs with a more balanced inhibition of reuptake of NA and 5-HT tend to be more effective in treating neuropathic pain than those that are much more potent at inhibiting reuptake of NA than of 5-HT [18]. In addition, although their best known neurochemical action is inhibition of reuptake of biogenic amines, the TCAs can affect other systems directly or indirectly, and their effectiveness may be the result of a variety of mechanisms, including alteration of 5-HT binding to its receptors, changing NA's interaction with α_2-adrenoreceptors, modification of opioid receptor density in the brain, binding to NMDA and/or AMPA glutamate receptors, blockade of sodium channels, inhibition of adenosine uptake in the periphery, and indirect dopaminergic actions [18, 67]. In addition, the atypical antidepressant bupropion (an NA and dopamine reuptake inhibitor) has been shown, as mentioned previously in this chapter, to be effective in neuropathic pain, suggesting that dopamine may also play an important role in antinociception [79].

The differences between the effectiveness of antidepressants based on their ability to inhibit reuptake of NA and 5-HT (discussed immediately above) and the reported effectiveness of bupropion in treating neuropathic pain [81], combined with the proposed indirect actions of TCAs on the dopaminergic system [18], has led to the idea that treatment of chronic pain and its depressive symptoms may be better accomplished with the targeting of multiple monoamine neurotransmitter systems. The recent development of triple reuptake inhibitors (which inhibit NA, 5-HT and dopamine reuptake) permits testing of this idea, and some triple reuptake inhibitors are currently undergoing preclinical investigations for use in pain and depression treatments [133–135].

Conclusion

There is a high degree of comorbidity between pain and depression which often complicates treatment of both conditions. The biological bases of this comorbidity are not well understood, and studies are complicated by interactions among various neurotransmitters and neuromodulators (e.g. NA, 5-HT, GABA, glutamate, neuroactive steroids, endogenous opioids, substance P), multiple receptors for the substances involved, and problems with translating animal model data to situations with humans. However, exciting recent advances in understanding the possible roles of biogenic amine and amino acid neurotransmitters in both depression and chronic pain through neurochemical, anatomical, molecular biological, pharmacological and behavioral studies as well as our better understanding of which antidepressants are most effective clinically in treating certain pain syndromes should result in more focused and effective treatment of co-occuring pain and depression in future.

Acknowledgements

Research funds have been provided by CIHR, the MS Society of Canada, the University of Alberta, The Freda and Joseph Berger Fund, and the Donald and Nancy Cranston Fund. The authors are grateful to Karyn Crawford for her expert secretarial assistance.

References

1 Breivik H, et al: Survey of chronic pain in Europe: prevalence, impact on daily life, and treatment. Eur J Pain 2006;10:287–333.
2 Elliott AM, et al: The epidemiology of chronic pain in the community. Lancet 1999;354:1248–1252.
3 Annagür BB, Uguz F, Apiliogullaro S, Kara I, Gunduz S: Psychiatric disorders and association with quality of sleep and quality of life in patients with chronic pain: a SCID-based study. Pain Med 2014;15:772–784.
4 Moriarty O, McGuire BE, Finn DP: The effect of pain on cognitive function: a review of clinical and preclinical research. Prog Neurobiol 2011;93:285–404.
5 Skaer TL: Fibromyalgia: disease synopsis, medication cost effectiveness and economic burden. Pharmacoeconomics 2014;32:457–466.
6 Van Velzen GA, Perez RS, van Gestel MA, Huygen FJ, van Kleef M, van Eijs F, Dahan A, van Hilten JJ, Marinus J: Health-related quality of life in 975 patients with complex regional pain syndrome type 1. Pain 2014;155:629–634.
7 Bair MJ, Kroenke K, Sutherland JM, McCoy KD, Harris H, McHorney CA: Effects of depression and pain severity on satisfaction in medical outpatients: analysis of the Medical Outcomes Study. J Rehabil Res Dev 2007;44:143–152.
8 Gureje O, Von Korff M, Kola L, Alonso J: The relation between multiple pains and mental disorders: results from the World Mental Health Surveys. Pain 2008;135:82–91.
9 Means-Christensen AJ, Roy-Byrne PP, Sherbourne CD, Craske MG, Stein MB: Relationships among pain, anxiety, and depression in primary care. Depress Anxiety 2008;25:593–600.
10 Karp JF, Scott J, Houch P, Reynolds CF 3rd, Kupfer DJ, Frank E: Pain predicts longer time to remission during treatment of recurrent depression. J Clin Psychiatry 2005;66:591–597.
11 Woo AKM: Depression and anxiety in pain. Rev Pain 2010;4:8–12.
12 Yalcin I, Barthas F, Barrot M: Emotional consequences of neuropathic pain: insight from preclinical studies. Neurosci Biobehav Rev 2014;47:154–164.
13 Arnow BA, Hunkeler EM, Blasey CM, Lee J, Constantino MJ, Fireman B, Kraemer HC, Dea R, Robinson R, Hayward C: Comorbid depression, chronic pain, and disability in primary care. Psychosom Med 2006;68:262–268.
14 Ohayon MM: Specific characteristics of the pain/depression association in the general population. J Clin Psychiatry 2004;65(suppl 12):5–9.
15 Lindsay PG, Wyckoff M: The depression-pain syndrome and its response to antidepressants. Psychosomatics 1981;22:571–573, 576–577.
16 Goldenburg DL: Pain/depression dyad: a key to better understanding and treatment of function somatic syndromes. Am J Med 2010;123:675–682.

17 Alev L, Lenox-Smith A, Altin M, Duenas H: A review of the serotonin-norepinephrine reuptake inhibitors: pharmacologic aspects and clinical implications for treatment of major depressive disorder and associated painful physical symptoms. Open J Depress 2013;2:54–63.
18 Sindrup SH, Otto M, Finnerup NB, Jensen TS: Antidepressants in the treatment of neuropathic pain. Basic Clin Pharmacol Toxicol 2005;96:399–409.
19 Verdu B, Decosterd I, Buclin T, Stiefel F, Berney A: Antidepressants for the treatment of chronic pain. Drugs 2008;68:2611–2632.
20 Briley M, Moret C: Treatment of comorbid pain with serotonin norepinephrine reuptake inhibitors. CNS Spectr 2008;13(7 suppl 11):22–26.
21 Mease PJ: Further strategies for treating fibromyalgia: the role of serotonin and norepinephrine reuptake inhibitors. Am J Med 2009;122(12 suppl):S44–S55.
22 Attal N, Cruccu G, Baron R, Haanpää M, Hansson P, Jensen TS, Nurmikko T; European Federation of Neurological Societies: EFNS guidelines on the pharmacological treatment of neuropathic pain: 2010 revision. Eur J Neurol 2010;17:1113–1188.
23 Saarto T, Wiffen PJ: Antidepressants for neuropathic pain: a Cochrane review. J Neurol Neurosurg Psychiatry 2010;81:1372–1373.
24 Matsuzawa-Yanagida K, Narita M, Nakajima M, Kuzumaki N, Nikura K, Nozaki H, Takagi T, Tamai E, Hareyama N, Terada M, Yamazaki M, Suzuki T: Usefulness of antidepressants for improving the neuropathic pain-like state and pain-induced anxiety through actions at different brain sites. Neuropsychopharmacology 2008;33:1952–1965.
25 Mico JA, Ardid D, Berrocosso E, Eschalier A: Antidepressants and pain. Trends Pharmacol Sci 2006;27:348–354.
26 Jann MW, Slade JH: Antidepressant agents for the treatment of chronic pain and depression. Pharmacotherapy 2007;27:1571–1587.
27 Schildkraut JJ: The catecholamine hypothesis of affective disorders: a review of supporting evidence. Am J Psychiatry 1965;122:509–522.
28 Lapin IP, Oxenkrug GF: Intensification of the central serotoninergic processes as a possible determinant of the thymoleptic effect. Lancet 1969;1:132–136.
29 Krishnan V, Nestler EJ: The molecular neurobiology of depression. Nature 2008;455:894–902.
30 Goldberg JS, Bell CE Jr, Pollard DA: Revisiting the monoamine hypothesis of depression: a new perspective. Perspect Med Chem 2014;6:1–8.
31 Paul IA, Skolnick P: Glutamate and depression: clinical and preclinical studies. Ann N Y Acad Sci 2003;1003:250–272.
32 Mitchell D, Baker GB: An update on the role of glutamate in the pathophysiology of depression. Acta Psychiatr Scand 2010;122:192–210.
33 Mohler H: The GABA system in anxiety and depression and its therapeutic potential. Neuropharmacology 2012;62:42–53.
34 Hashimoto K: The role of glutamate in the action of antidepressants. Prog Neuropsychopharmacol Biol Psychiatry 2011;35:156–158.
35 Mann J, Malone KM: Cerebospinal fluid amines and higher-lethality suicide attempts in depressed inpatients. Biol Psychiatry 1997;41:162–171.
36 Owens MJ, Nemeroff CB: Role of serotonin in the pathophysiology of depression: focus on the serotonin transporter. Clin Chem 1994;40:288–295.
37 Julius D, Basbaum AI: Molecular mechanisms of nociception. Nature 2001;413:203–210.
38 Moore KA, Kohno T, Karchewski LA, Scholz J, Baba H, Woolf CJ: Partial peripheral nerve injury promotes a selective loss of GABAergic inhibition in the superficial dorsal horn of the spinal cord. J Neurosci 2002;22:6724–6731.
39 Bazzichi L, Giannaccini G, Betti L, Mascia G, Fabbrini L, Italiani P, De Feo F, Giuliano T, Giacomelli C, Rossi A, Lucacchini A, Bombardieri S: Alteration of serotonin transporter density and activity in fibromyalgia. Arthritis Res Ther 2006;8:R99.
40 Gwak YS, Hulsebosch CE: GABA and central neuropathic pain following spinal cord injury. Neuropharmacology 2011;60:799–808.
41 Larsson M, Broman J: Synaptic plasticity and pain: role of ionotropic glutamate receptors. Neuroscientist 2010;17:256–273.
42 Le Coz GM, Fiatte C, Anton F, Hanesch U: Differential neuropathic pain sensitivity and expression of spinal mediators in Lewis and Fischer 344 rats. BMC Neurosci 2014;15:35.
43 Bravo L, Torres-Sanchez S, Alba-Delgado C, Mico JA, Berrocoso E: Pain exacerbates chronic mild stress-induced changes in noradrenergic transmission in rats. Neuropharmacology 2014;24:996–1003.
44 Rodríguez-Gaztelumendi A, Rojo ML, Pazos A, Díaz Á: An altered spinal serotonergic system contributes to increase thermal nociception in an animal model of depression. Exp Brain Res 2014;232:1793–1803.
45 Delgado PL: Common pathways of depression and pain. J Clin Psychiatry 2004;65(suppl 12):16–19.
46 Millan MJ: Descending control of pain. Prog Neurobiol 2002;66:355–474.
47 Rainville P, Duncan GH, Price DD, Carrier B, Bushnell MC: Pain affect encoded in human anterior cingulate but not somatosensory cortex. Science 1997;277:968–971.
48 Bär K, Wagner G, Kosche M, Boettger S, Boettger MK, Schlosser R, Sauer H: Increased prefrontal activation during pain perception in major depression. Biol Psychiatry 2007;62:1281–1287.

49 Giesecke T, Gracely RH, Williams DA, Geisser ME, Petzke FW, Clauw DJ: The relationship between depression, clinical pain, and experimental pain in a chronic pain cohort. Arthritis Rheum 2005;52:1577–1584.

50 Shackman AJ, Salomons TV, Slagter HA, Fox AS, Winter JJ, Davidson RJ: The integration of negative affect, pain and cognitive control in the cingulate cortex. Nat Rev Neurosci 2011;12:154–167.

51 Drevets WC: Neuroimaging and neuropathological studies of depression: implications for the cognitive-emotional features of mood disorders. Curr Opin Neurobiol 2001;11:240–249.

52 Alba-Delgado C, Llorca-Torralba M, Horrillo I, Ortega JE, Mico JA, Sanchez-Blazquez P, Meana JJ, Berrocoso E: Chronic pain leads to concomitant noradrenergic impairment and mood disorders. Biol Psychiatry 2013;73:54–62.

53 Ordway GA, Streator Smith K, Haycock JW: Elevated tyrosine hydroxylase in the locus coeruleus of suicide victims. J Neurochem 1994;62:680–685.

54 Ordway GA, Widdowson PS, Streator Smith K, Halaris A: Agonist binding to alpha 2-adrenoceptors is elevated in the locus coeruleus from victims of suicide. J Neurochem 1994;63:617–624.

55 Zhu MY, et al: Elevated levels of tyrosine hydroxylase in the locus coeruleus in major depression. Biol Psychiatry 2009;46:1275–1286.

56 Martins I, de Vries MG, Teixeira-Pinto A, Fadel J, Wilson SP, Westerink BH, Tavares I: Noradrenaline increases pain facilitation from the brain during inflammatory pain. Neuropharmacology 2013;71: 299–307.

57 Bardin L: The complex role of serotonin and 5-HT receptors in chronic pain. Behav Pharmacol 2011;22: 390–404.

58 Lin SY, Chang WJ, Lin CS, et al: Serotonin receptor 5-HT2B mediates serotonin-induced mechanical hyperalgesia. Neuroscience 2011;31:1410–1418.

59 Suzuki T, Amata M, Sakaue G, Nishimura S, Inoue T, Shibata M, Mashimo K: Experimental neuropathy in mice is associated with delayed behavioral changes related to anxiety and depression. Anesth Analg 2007;104:1570–1577.

60 Vanegas H, Schaible HG: Descending control of persistent pain: inhibitory or facilitatory? Brain Res 2004;46:295–309.

61 Lopez-Garcia JA: Serotonergic modulation of spinal sensory circuits. Curr Top Med Chem 2006;6:1987–1996.

62 Mickle A, Kannamapalli P, Bruckert M, Miranda A, Banerjee B, Sengupta JN: Pronociceptive effect of 5-HT_{1A} receptor agonist on visceral pain involves spinal N-methyl-D-aspartate (NMDA) receptor. Neuroscience 2012;219:243–254.

63 Miranda A, Peles S, McLean PG, Sengupta JN: Effects of the 5-HT_3 receptor antagonist, alosteron, in a rat model of somatic and visceral hyperalgesia. Pain 2006;126:54–63.

64 Sengupta JN, Mickle A, Kannampalli P, Spruell R, McRorie J, Shaker R, Miranda A: Visceral analgesic effect of 5-HT_4 receptor agonist in rats involves the rostroventral medulla (RVM). Neuropharmacology 2014;79:345–358.

65 Dworkin RH, O'Connor AB, Backonja M, Farrar JT, Finnerup NB, Jensen TS, Kalso EA, Loeser JD, Miaskowski C, Nurmikko TJ, Portenoy RK, Rice AS, Stacey BR, Treede RD, Turk DC, Wallace MS: Pharmacologic management of neuropathic pain: evidence-based recommendations. Pain 2007;132: 237–251.

66 McGleane G: Antidepressants as analgesics. CNS Drugs 2008;22:139–156.

67 Mika J, Zychowska M, Makuch W, Rojewska E, Przewlocka B: Neuronal and immunological basis of action of antidepressants in chronic pain – clinical and experimental studies. Pharmacol Rep 2013;65:1611–1621.

68 Davidson J: Monoamine oxidase inhibitors in patients with chronic pain. Arch Gen Psychiatry 1985; 42:635–636.

69 Turkewitz LJ, Casaly JS, Dawson GA, Wirth O: Phenelzine therapy for headache patients with concomitant depression and anxiety. Headache 1992;32: 203–207.

70 Musgrave T, Benson C, Wong G, Browne I, Tenorio G, Rauw G, Baker GB, Kerr BJ: The MAO inhibitor phenelzine improves functional outcomes in mice with experimental autoimmune encephalomyelitis (EAE). Brain Behav Immun 2011;25:1677–1688.

71 Benson CA, Wong G, Tenorio G, Baker GB, Kerr BJ: The MAO inhibitor phenelzine can improve functional outcomes in mice with established clinical signs in experimental autoimmune encephalomyelitis (EAE). Behav Brain Res 2013;252:302–311.

72 Baker GB, Wong JTF, Yeung JM, Coutts RT: Effects of the antidepressant phenelzine on brain levels of γ-aminobutyric acid (GABA). J Affect Disord 1991; 21:207–211.

73 Bai L, Wang W, Dong YL, Wang W, Huang J, Wang XY, Wang LY, Li YQ, Wu SX: Attenuation of mouse somatic and emotional inflammatory pain by hydralazine through scavenging acrolein and inhibiting neuronal activation. Pain Physician 2012;15: 311–326.

74 Due MR, Park J, Zheng L, Walls M, Allette YM, White FA, Shi R: Acrolein involvement in sensory and behavioral hypersensitivity following spinal cord injury in rats. J Neurochem 2014;128:776–786.

75 Wood PL, Khan MA, Moskal JR, Todd KG, Tanay VA, Baker G: Aldehyde load in ischemia-reperfusion brain injury: neuroprotection by neutralization of reactive aldehydes with phenelzine. Brain Res 2006; 1122:184–190.
76 Wood PB: Role of central dopamine in pain and analgesia. Expert Rev Neurother 2008;8:781–797.
77 Potvin S, Grignon S, Marchand S: Human evidence of a supra-spinal modulating role of dopamine on pain perception. Synapse 2009;63:390–402.
78 Jarcho JM, Mayer EA, Jiang ZK, Feier NA, London ED: Pain, affective symptoms, and cognitive deficits in patients with cerebral dopamine dysfunction. Pain 2012;153:744–754.
79 Laruelle M: The role of endogenous sensitization in the pathophysiology of schizophrenia: implications from recent brain imaging studies. Brain Res Rev 2000;31:371–384.
80 Hagelberg N, Jaaskelainen SK, Martikainen IK, et al: Striatal dopamine D_2 receptors in modulation of pain in humans: a review. Eur J Pharmacol 2004;500: 187–192.
81 Semenchuk MR, Sherman S, Davis B: Double-blind, randomized trial of bupropion SR for the treatment of neuropathic pain. Neurology 2001;57:1582–1588.
82 Green AR, Hainsworth AH, Jackson DM: GABA potentiation: a logical pharmacological approach for the treatment of acute ischaemic stroke. Neuropharmacology 2000;39:1483–1494.
83 Luscher B, Fuchs T: GABAergic control of depression-related brain states. Adv Pharmacol 2015;73: 97–144.
84 Kendell SF, Krystal JH, Sanacora G: GABA and glutamate systems as therapeutic targets in depression and mood disorders. Exp Opin Ther Targets 2005;9: 153–168.
85 Petroff OA: GABA and glutamate in the human brain. Neuroscientist 2002;8:562–573.
86 El-Ansary A, Al-Ayadhi L: GABAergic/glutamatergic imbalance relative to excessive neuroinflammation in autism spectrum disorders. J Neuroinflammation 2014;11:189–200.
87 Pehrson AL, Sanchez C: Altered gamma-aminobutryic acid neurotransmission in major depressive disorder: a critical review of the supporting evidence and the influence of serotonergic antidepressants. Drug Des Devel Ther 2015;9:603–624.
88 Polgar E, Watanabe M, Hartmann B, Grant SG, Todd AJ: Expression of AMPA receptor subunits at synapses in laminae I-III of the rodent spinal dorsal horn. Mol Pain 2008;4:5.
89 Zeilhofer HU, Möhler H, Di Lio A: GABAergic analgesia: new insights from mutant mice and subtype-selective agonists. Cell 2009;30:397–402.
90 Goudet C, Magnaghi V, Landry M, Nagy F, Gereau RW 4th, Pin JP: Metabotropic receptors for glutamate and GABA in pain. Brain Res Rev 2009;60: 43–56.
91 Kannampalli P, Sengupta JN: Role of principal ionotropic and metabotropic receptors in visceral pain. J Neurogastroenterol Motil 2015;21:147–158.
92 Tanay VM, Greenshaw AJ, Baker GB, Bateson AN: Common effects of chronically administered antipanic drugs on brainstem $GABA_A$ receptor subunit gene expression. Mol Psychiatry 2001;6:404–412.
93 Tzellos TG, Papazisis G, Amaniti E, Kouvelas D: Efficacy of pregabalin and gabapentin for neuropathic pain in spinal-cord injury: an evidence-based evaluation of the literature. Eur J Clin Pharmacol 2008;64:851–858.
94 Moulin DE, Boulanger A, Clark AJ, Clarke H, Dao T, Finley GA, Furlan A, Gilron I, Gordon A, Morley-Forster PK, Sessle BJ, Squire P, Stinson J, Taenzer P, Velly A, Ware MA, Weinberg EL, Williamson OD: Pharmacological management of chronic neuropathic pain: revised consensus statement from the Canadian Pain Society. Pain Res Manag 2014;19:328–335.
95 Taylor CP: Gapapentin-mechanisms of action; in Levy RH, Mattson RH, Meldrum BS (eds): Antiepileptic Drugs. New York, Raven Press, 1995, pp 829–841.
96 Tanabe M, Takasu K, Takeuchi Y, Ono H: Pain relief by gabapentin and pregabalin via supraspinal mechanisms after peripheral nerve injury. J Neurosci 2008;86:3258–3264.
97 Field MJ, Cox PJ, Stott E, Melrose H, Offord J, Su TZ, Bramwell S, Corradini L, England S, Winks J, Kinloch RA, Hendrich J, Dolphin AC, Webb T, Williams D: Identification of the α_2-δ-1 subunit of voltage-dependent calcium channels as a molecular target for pain mediating the analgesic action of pregabalin. Proc Natl Acad Sci U S A 2006;103: 17537–17542.
98 Hayashida KI, Degoes S, Curry R, Eisenach JC: Gabapentin activates spinal noradrenergic activity in rats and humans and reduces hypersensitivity after surgery. Anesthesiology 2007;106:557–562.
99 Yoshizumi M, Parker RA, Eisenach JC, Hayshida K: Gabapentin inhibits γ-amino butyric acid release in the locus coeruleus but not in the spinal dorsal horn after peripheral nerve injury in rats. Anesthesiology 2012;116:1347–1353.
100 Suto T, Severino AL, Eisenach JC, Hayashida K-I: Gabapentin increases extracellular glutamergic level in the locus coeruleus via astroglial glutamate transporter-dependent mechanisms. Neuropharmacology 2014;81:95–100.

101 Coderre TJ, Melzack R: The contribution of excitatory amino acids to central sensitization and persistent nociception after formalin-induced tissue injury. J Neurosci 1992;12:3665–3670.
102 Coderre TJ: The role of excitatory amino acid receptors and intracellular messengers in persistent nociception after tissue injury in rats. Mol Neurobiol 1993;7:229–246.
103 Bleakman D, Alt A, Nisenbaum ES: Glutamate receptors and pain. Semin Cell Dev Biol 2006;17: 592–604.
104 Chizh BA, Headley PM: NMDA antagonists and neuropathic pain – multiple drug targets and multiple uses. Curr Pharm Des 2005;11:2977–2994.
105 Goffer Y, Xu D, Eberle SE, D'amour J, Lee M, Tukey D, Froemke RC, Ziff EB, Wang J: Calcium-permeable AMPA receptors in the nucleus accumbens regulate depression-like behaviours in the chronic neuropathic pain state. J Neurosci 2013;33: 19034–19044.
106 Naughton M, Clarke G, O'Leary OF, Cryan JF, Dinan TG: A review of ketamine in affective disorders: current evidence of clinical efficacy, limitations of use and pre-clinical evidence on proposed mechanisms of action. J Affect Disord 2014;156: 24–35.
107 Iadorola ND, Niciu MJ, Richards AM, Vande Voort JL, Ballard ED, Lundin NB, Nugent AC, Machado-Vieira R, Zarate CA Jr: Ketamine and other N-methyl-D-aspartate receptor antagonists in the treatment of depression: a perspective review. Ther Adv Chronic Dis 2015;6:97–114.
108 Niesters M, Martini C, Dahan A: Ketamine for chronic pain: risks and benefits. Br J Clin Pharmacol 2014;77:357–367.
109 Beaulieu EE: Neurosteroids: of the nervous system, by the nervous system, for the nervous system. Rec Prog Horm Res 1997;52:1–32.
110 Mellon SH, Griffin LD, Compagnone NA: Biosynthesis and action of neurosteroids. Brain Res Rev 2001;37:3–12.
111 Dubrovsky BO: Steroids, neuroactive steroids and neurosteroids in psychopathology. Prog Neuropsychopharmacol Biol Psychiatry 2005;29:169–192.
112 MacKenzie EM, Odontiadis J, Le Melledo J-M, Prior T, Baker GB: The relevance of neuroactive steroids in schizophrenia, depression, and anxiety disorders. Cell Mol Neurobiol 2007;27:541–574.
113 Schule C, Eser D, Baghai TC, Nothdurfter C, Kessler JS, Rupprecht R: Neuroactive steroids in affective disorders: target for novel antidepressant or anxiolytic drugs? Neuroscience 2011;191:55–77.
114 Leonelli E, Bianchi R, Cavaletti G, et al: Progesterone and its derivatives are neuroprotective agents in experimental diabetic neuropathy: a multimodal analysis. Neuroscience 2007;4:1293–1304.
115 Menasch-Nyagan AG, Kibaly C, Schaeffer V, Venard C, Meyer L, Patte-Mensah C: Endogenous steroid production in the spinal cord and potential involvement in neuropathic pain modulation. J Steroid Biochem Mol Biol 2008;109:286–293.
116 Charlet A, Lasbennes F, Darbon P, Poisbeau P: Fast non-genomic effects of progesterone-derived neurosteroids on nociceptive thresholds and pain symptoms. Pain 2008;139:603–609.
117 Jevtovic-Todorovic V, Covey DF, Todorovic SM: Are neuroactive steroids promising therapeutic agents in the management of acute and chronic pain? Psychoneuroendocrinology 2009;34(S1): S178–S185.
118 Afrazi S, Esmaeili-Mahani S: Allopregananolone suppresses diabetes-induced neuropathic pain and motor deficit through inhibition of GABAA receptor down-regulation in the spinal cord of diabetic rats. Iran J Basic Med Sci 2014;17:312–317.
119 Melcangi RC, Panzica GC: Allopregnanolone: state of the art. Prog Neurobiol 2014;113:1–5.
120 Romeo E, Strohle A, Spalletta G, di Michele F, Hermann B, Holsboer F, Pasini A, Rupprecht R: Effects of antidepressant treatment on neuroactive steroids in major depression. Am J Psychiatry 1998; 155:910–913.
121 Nechmad A, Maayan R, Spivak B, Ramadan E, Poyurovsky M, Weizman A: Brain neurosteroid changes after paroxetine administration in mice. Eur J Neuropsychopharmacol 2003;13:327–332.
122 Uzunova V, Sheline Y, Davis JM, Rasmusson A, Uzunov DP, Costa E, Guidotti A: Increase in the cerebrospinal fluid content of neurosteroids in patients with unipolar major depression who are receiving fluoxetine or fluvoxamine. Proc Natl Acad Sci U S A 1998;95, 3239–3244.
123 Eser D, Schüle C, Baghai TC, Romeo E, Rupprecht R: Neuroactive steroids in depression and anxiety disorders: clinical studies. Neuroendocrinology 2006;84:244–254.
124 Pinna G, Costa E, Guidotti A: Fluoxetine and norfluoxetine stereospecifically and selectively increase brain neurosteroid content at doses that are inactive on 5-HT reuptake. Psychopharmacology 2006;186:362–372.
125 Patte-Mensah C, Meyer L, Taleb O, Mensah-Nyagan AG: Potential role of allopregnanolone for a safe and effective therapy of neuropathic pain. Prog Neurobiol 2013;113:70–78.
126 Smith AJ: The analgesic effects of selective serotonin reuptake inhibitors. J Psychopharmacol 1998;12:408–413.
127 Dhillon S: Duloxetine: a review of its use in the management of major depressive disorder in older adults. Drugs Aging 2013;30:59–79.

128 Gendreau RM, Thorn MD, Gendreau JF, Kranzier JD, Ribeiro S, Gracely RH, Williams DA, Mease PJ, McLean SA, Clauw DJ: Efficacy of milnacipran in patients with fibromyalgia. J Rheumatol 2005;32: 1975–1985.
129 Marks DM, Shah MJ, Patkar AA, Masand PS, Park GY, Pae CU: Serotonin-norepinephrine reuptake inhibitors for pain control: premise and promise. Curr Neuropharmacol 2009;7:331–336.
130 Manna V, Bolino F, Di CL: Chronic tension-type headache, mood depression and serotonin: therapeutic effects of fluvoxamine and mianserine. Headache 1994;34:44–49.
131 Mattia C, Paoletti F, Coluzzi F, Boanelli A: New antidepressants in the treatment of neuropathic pain. A review. Minerva Anestesiol 2002;68:105–114.
132 Stahl S, Grady MM, Moret C, Briley M: SNRIs: The pharmacology, clinical efficacy, and tolerability in comparison with other classes of antidepressants. CNS Spectr 2005;10:732–747.
133 Basile AS, Janowsky A, Golemiowska K, Kowalska M, Tam E, Beneveniste M, Popik P, Nikiforuk A, Krawczyk M, Nowak G, Krieter PA, Lippa AS, Skolnick P, Koustova E: Characterization of the antinociceptive actions of bicifadine in models of acute, persistent, and chronic pain. J Pharmacol Exp Ther 2007;321:1208–1225.
134 Hache G, Coudore F, Gardier AM, Guiard BP: Monoaminergic antidepressants in the relief of pain: potential therapeutic utility of triple reuptake inhibitors (TRIs). Pharmaceuticals 2011;4:285–342.
135 Hache G, Guiard BP, Nguyen TH, Quesseveur G, Gardier AM, Peters D, Munro G, Coudore F: Antinociceptive activity of the new triple reuptake inhibitor NS18283 in a mouse model of chemotherapy-induced neuropathic pain. Eur J Pain 2015;19: 322–333.

Prof. Glen Baker
Neuroscience and Mental Health Institute, University of Alberta
12-105B Clinical Sciences Building
Edmonton, AB T6G 2B7 (Canada)
E-Mail glen.baker@ualbert.ca

Finn DP, Leonard BE (eds): Pain in Psychiatric Disorders. Mod Trends Pharmacopsychiatry.
Basel, Karger, 2015, vol 30, pp 80–93 (DOI: 10.1159/000435934)

Supraspinal Transient Receptor Potential Subfamily V Member 1 (TRPV1) in Pain and Psychiatric Disorders

Manish K. Madasu[a, c] · Michelle Roche[b, c] · David P. Finn[a, c]

[a]Pharmacology and Therapeutics, and [b]Physiology, School of Medicine, and [c]Galway Neuroscience Centre and Centre for Pain Research, NCBES, National University of Ireland, Galway, Ireland

Abstract

The transient receptor potential subfamily V member 1 (TRPV1) belongs to the diverse transient receptor potential (TRP) family of cation channels. It was first characterized in primary afferent fibres as a receptor for capsaicin. Peripheral TRPV1 has a very well-described role in nociception. However, TRPV1 is now recognized to have a broader distribution and function, with supraspinal/brain TRPV1 known to modulate pain processing. Recently, studies employing histological, genetic and pharmacological approaches have provided evidence that supraspinal TRPV1 also modulates brain neurobiology and behaviours related to anxiety, depression and schizophrenia. Key brain regions involved in TRPV1-mediated modulation of pain and affect include the periaqueductal grey, hippocampus and medial prefrontal cortex. Thus, TRPV1 in the brain is emerging as an important molecular substrate which is dually implicated in both pain and psychiatric disorders, and represents a novel therapeutic target for these conditions and their comorbidity.

Introduction to Transient Receptor Potential Subfamily V Member 1

Although first identified as the receptor for capsaicin, transient receptor potential subfamily V member 1 (TRPV1) can also be activated endogenously by voltage, noxious heat (>42°C), low pH and lipoxygenase products. Capsaicin is the naturally occurring pungent constituent of capsicum peppers used in many hot/spicy foods. TRPV1 expressed on primary afferent neurones that detect and encode noxious stimuli can be activated by capsaicin, resulting in neuronal excitation and release of local inflammatory mediators [1]. Endocannabinoids/endovanilloids, including anandamide and N-arachidonoyl dopamine, are endogenous ligands that can also activate TRPV1. TRPV1 nonselectively gates cations; however, channel activation results in a 10-fold higher preference for calcium [2–4].

TRPV1 is comprised of six transmembrane domains and intracellular N- and C-termini. The N-terminal tail contains numerous phosphorylation sites and ankyrin repeats that serve as binding sites for calmodulin and adenosine triphosphate [5]. The C-terminal tail contains a TRP domain as well as binding sites for both calmodulin and phosphatidylinositol (4,5)-bisphosphate, an endogenous TRPV1 inhibitor [6]. Agonist binding and receptor activation can occur intracellularly, as lipophilic capsaicin readily crosses the membrane to bind several sites on TRPV1 [7]. TRPV1 is expressed highly in the dorsal root ganglia of C and Aδ fibres. In these fibres, TRPV1 activation leads to increases in intracellular calcium levels which in turn induces the release of neuropeptides (calcitonin gene-related peptide and substance P) in the dorsal horn of the spinal cord [8].

Studies in rat and primate brains have shown that TRPV1 is widely expressed throughout the neuroaxis, including the cortex, hippocampus, basal ganglia, cerebellum and olfactory bulb, as well as in the mesencephalon and hindbrain [9, 10]. Studies of the distribution of TRPV1 in human brain have been more restricted, but a post-mortem study has shown that TRPV1 receptors have been found in the third and fifth layers of the human parietal cortex [9]. However, overall TRPV1 expression in the central nervous system (CNS) is considerably lower than in the dorsal root ganglia [10–13]. Indeed, some studies have failed to detect the presence of TRPV1 in the CNS [1, 14–16] possibly due to complexity in genes and strain-related variations [11, 17]. A sophisticated gene strategy where the TRPV1 gene was targeted by attaching two reporters, PLAP (placental alkaline phosphatase) and nlacZ (nuclear lacZ), onto the TRPV1 promoter gene and creating a specific line of mice (TRPV1$^{\text{PLAP-nlacZ}}$) was used to confirm TRPV1 expression in the CNS. This study reported that TRPV1 expression in the CNS is limited to certain brain regions and low when compared to expression in dorsal root ganglia [12]. This restricted expression of TRPV1 in the CNS was confirmed by in situ hybridization experiments in rat, monkey and human brains [12]. However, a number of recent pharmacological, genetic, radioligand-binding and immunohistochemical studies suggest widespread distribution and functionality of TRPV1 across the CNS [11, 12, 18–20].

Role of Transient Receptor Potential Subfamily V Member 1 in the Brain

Pain

Acute Pain

The periaqueductal grey-rostral ventromedial medulla (PAG-RVM) pathway is very important in pain processing and modulation. PAG-mediated antinociception involves the recruitment of pain-modulating RVM neurons via the descending pain pathway [21]. The existence of TRPV1 in the midbrain has been demonstrated by immunohistochemistry [9, 22, 23], in situ hybridization [9], binding of the TRPV1-selective radioligand [3H]-resiniferatoxin (RTX) [24] and gene reporter studies [12]. The RVM contains three different types of pain-responsive neurons: 'neutral cells', which show no modification in spontaneous activity associated with nociceptive stimulation; ON cells, which show a burst of activity before withdrawal reflexes, and OFF cells, which are inhibited just before withdrawal reflexes [25–27]. Capsaicin, when injected into the dorsolateral periaqueductal grey (dlPAG), increased the latency of nociceptive responding to noxious heat, indicating that stimulation of TRPV1 within the descending inhibitory pain pathway can cause antinociception [28]. Microinjection of capsaicin into the ventrolateral periaqueductal grey (vlPAG) increased the threshold of thermal pain sensitivity in rats [29]. Opposite effects were found with 5-iodoresiniferatoxin (I-RTX), a selective TRPV1 antagonist that facilitated nociceptive responses and, at an inactive dose, abolished capsaicin-mediated antinociception, implying that the effect of capsaicin is mediated by TRPV1 in the vlPAG [29]. The antinociceptive effect of intra-vlPAG capsaicin was accompanied by an increase in glutamate release in the RVM as measured by in vivo microdialysis, which was also blocked by an inactive dose of I-RTX. The TRPV1 antagonist itself reduced the release of glutamate, thus suggesting that vlPAG TRPV1 tonically stimulates glutamatergic output to the RVM with a concomitant inhibition of nociception [29]. Hyperalgesia or analgesia have been observed following intra-vlPAG administration of the fatty acid amide hydrolase (FAAH) inhibitor URB597 depending on whether vlPAG cannabinoid receptors or TRPV1 have been activated [30]. It was proposed that anandamide-mediated activation of TRPV1 leads to analgesia, while hyperalgesia may be due to increases in vlPAG 2-arachidonoylglycerol leading to CB_1 receptor stimulation, which in turn leads to inhibition of the antinociceptive PAG-RVM descending pathway [30].

The ON and OFF neurons in the RVM have been shown to respond to capsaicin administered into the PAG [29–31]. Intra-dlPAG microinjection of capsaicin is followed by a decrease in the tail flick-related ON cell burst activity and an increase in tail flick latency [31]. Later on, due to desensitization of the receptor (due to prolonged exposure to capsaicin), antinociception correlating with increased OFF cell activity was reported [31]. Similarly, Starowicz et al. [29] have shown that intra-vlPAG administration of capsaicin caused a decrease in the firing activity of RVM ON cells, and an increase in the firing of the OFF cells. Moreover, microinjections of capsaicin into the vlPAG have also been shown to increase withdrawal latency in the rat hot-plate test, with evidence that activation of TRPV1 in the vlPAG induces antinociception via mGlu receptor-mediated 2-arachidonoylglycerol retrograde signalling in the RVM [32]. Intra-vlPAG administration of the FAAH inhibitor URB597, which is known to enhance endogenous anandamide levels, stimulated OFF cell activity in the RVM and inhibited ON cell activity [30]. This effect on RVM activity was abolished by intra-vlPAG administration of the TRPV1 antagonist capsazepine, suggesting that FAAH substrates (likely anandamide) activate TRPV1 on vlPAG neurons, with projections from the PAG to the RVM mediating the subsequent stimulation of RVM OFF cells and inhibition of ON cells. De Novellis et al. [33] administered N-arachidonoyl-serotonin (AA-5-HT), a compound with a dual ability to inhibit FAAH and block TRPV1, into the vlPAG, and measured endocannabinoid levels, RVM ON and OFF cell activities, thermal nociception in the tail flick test and formalin-induced nociceptive behaviour. They found that AA-5-HT increased anandamide levels in the vlPAG and had antinociceptive effects in both the tail flick and formalin tests. Moreover, intra-vlPAG administration of AA-5-HT depressed the activity of both OFF cell and ON cells in the RVM. These effects of AA-5-HT were similar to those seen following coadministration of the FAAH inhibitor URB597 and the selective TRPV1 antagonist I-RTX into the vlPAG [33]. The FAAH substrate, palmitoylethanolamide (PEA), when microinjected into the vlPAG of rats, was antinociceptive in the tail flick test, concomitantly decreasing the ongoing activity of the OFF cells in the RVM and increasing the latency of tail flick-evoked onset of ON cell activity [34]. These latter effects on RVM cell activity were blocked by the TRPV1 antagonist I-RTX, suggesting that TRPV1 modulates PEA-induced effects within the PAG-RVM circuitry [34]. PEA does not directly bind to TRPV1, but through substrate competition at FAAH it can indirectly elevate AEA levels, which in turn may bind to TRPV1 and induce antinociception.

Chronic Pain

Intracerebroventricular administration of the TRPV1 receptor antagonist (1-[3-(trifluoromethyl)pyridin-2-yl]-N-[4-(trifluoromethyl sulfonyl)phenyl]-1,2,3,6-tetrahydropyridine-4-carboxamide) A-784168 significantly reduced weight bearing in the sodium monoiodoacetate model of osteoarthritis and reduced complete Freund's adjuvant-induced chronic inflammatory thermal hyperalgesia, suggesting that TRPV1 in the brain plays a key role in chronic inflammatory pain [35]. Furthermore, intracerebroventricular administration of the plant-derived alkaloid (-)cassine prevented mechanical hyperalgesia induced by carrageenan, an effect mediated by TRPV1 receptors [36]. Site-specific delivery of the dual TRPV1 and FAAH blocker, AA-5-HT, into the prelimbic-infralimbic cortex significantly decreased mechanical allodynia in the spared nerve injury model of neuropathic pain in mice [37]. Similarly, intracortical (prelimbic-infralimbic cortex) administration of AA-5-HT was more effective in reducing spared nerve injury-induced mechanical allodynia than I-RTX [38]. In addition, evidence suggests reciprocal alterations in TRPV1 and the CB_1 receptor mediate visceral hyperalgesia in water avoidance stressed rats [39]. Thus, while fewer studies have investigated the role of supraspinal TRPV1 in animal models of chronic pain versus acute pain, these studies together suggest involvement of supraspinal TRPV1 in the development and/or maintenance of both chronic inflammatory and neuropathic pain. Table 1 provides a summary of studies to date that have investigated the effects of intracerebral administration of TRPV1 ligands on nociceptive behaviour in animal models of acute and chronic pain.

Table 1. Effects of pharmacological modulation of supraspinal TRPV1 in animal models of pain

Drug and dose	Route of admin./target region in CNS	Test/effect on behaviour	Animal	Reference
Capsaicin (1, 3, 6 nmol/rat)	i.c. (dlPAG)	Antinociceptive effect/thermosensitivity test	Rat	[28]
Capsazepine (6 nmol/rat)	i.c. (dlPAG)	Blocked the capsaicin-induced antinociception/thermosensitivity test	Rat	[28]
Capsaicin (6 nmol)	i.c. (vlPAG)	Antinociceptive effect/thermosensitivity test	Rat	[30]
Capsaicin (3 and 6 nmol/rat)	i.c. (vlPAG)	Antinociceptive effect/RVM extracellular recordings and tail flick test	Rat	[29]
I-RTX (0.5 and 1 nmol/rat)	i.c. (vlPAG)	Blocked the antinociceptive effect of capsaicin/RVM extracellular recordings and tail flick test	Rat	[29]
Capsaicin (10 nmol)	i.c. (dPAG)	Initially produced hyperalgesia followed by analgesia/RVM extracellular recordings and tail flick test	Rat	[31]
Capsazepine (10 nmol)	i.c. (dPAG)	Blocked the hyperalgesic effect of capsaicin/RVM extracellular recordings and tail flick test	Rat	[31]
Capsaicin (6 nmol)	i.c. (vlPAG)	Antinociceptive effect/hot-plate test	Rat	[32]
SB366791 (50 nmol)	i.c. (vlPAG)	Abolished antinociceptive effect/hot-plate test	Rat	[32]
AA-5-HT (0.1, 0.2, 0.5 nmol)	i.c. (vlPAG)	Pronociceptive at lower doses and antinociceptive at higher doses/RVM extracellular recordings and tail flick test	Rat	[33]
I-RTX (0.5 nmol)	i.c. (vlPAG)	Blocked the AA-5-HT effects/RVM extracellular recordings and tail flick test	Rat	[33]
PEA (3 or 6 nmol)	i.c. (vlPAG)	Antinociceptive effect/RVM extracellular recordings and tail flick test	Rat	[34]
I-RTX (1 nmol)	i.c. (vlPAG)	Blocked the PEA-induced effects/RVM extracellular recordings and tail flick test	Rat	[34]
A-784168 (100 nmol)	i.c.v.	Reduced weight bearing/sodium monoiodoacetate model of osteoarthritis	Rat	[35]
A-784168 (100 nmol)	i.c.v.	Reduced CFA-induced chronic inflammatory thermal hyperalgesia/thermal test	Rat	[35]
Cassine (10 μg)	i.c.v.	Blocked mechanical hyperalgesia/carrageenan model	Rat	[36]
AA-5-HT (0.1, 0.25, 1 nmol), I-RTX (0.25, 0.5, 1 nmol)	Intracortical (PL-IL)	Reducing mechanical allodynia/SNI model	Rat	[37, 38]

i.c. = Intracerebral; i.c.v. = intracerebroventricular; SB366791 = N-(3-methoxyphenyl)-4-chlorocinnamide; A-784168 = 1-[3-(trifluoromethyl)pyridin-2-yl]-N-[4-(trifluoromethyl sulfonyl)phenyl]-1,2,3,6-tetrahydropyridine-4-carboxamide; PL-IL = prelimbic-Infralimbic cortex; SNI = spared nerve injury.

Psychiatric Disorders

Transient Receptor Potential Subfamily V Member 1 and Anxiety

Generalized Anxiety Disorder. The pharmacological studies summarized in table 2 indicate that TRPV1 in the dorsal periaqueductal grey (dPAG), the hippocampus, the medial prefrontal cortex and the basolateral amygdala modulates anxiety-related behaviour in the rat/mouse elevated plus maze (EPM). Systemic injection of TRPV1 antagonists (capsazepine and SB-366791) or agonist (Olvanil) has been shown to produce anxiolytic or anxiogenic effects, respectively, in the EPM [40, 41]. TRPV1 knockout (KO) mice exhibit an anxiolytic phenotype, suggesting that TRPV1 plays a role in anxiety-related behaviour [42]. Microinjections of the TRPV1 antagonists capsazepine or 6-iodo-nordihydrocapsaicin into the ventromedial prefrontal cortex had an anxiolytic effect in the EPM and Vogel conflict tests, and attenuated the expression of contextual fear conditioning in rats [43–45]. Similar to the ventromedial prefrontal cortex, blockade of TRPV1 in the ventral hippocampus and dlPAG also elicited anxiolytic effects in the EPM [46, 47].

Panic Disorder. The dlPAG is known for its role in coordinating freezing, fight and flight behaviours in threatening situations, such as the presence of a predator [50, 51]. Evidence suggests that electrical stimulation of the dlPAG in humans induces symptoms similar to a panic attack [52]. Corroborating this finding, neuroimaging studies have shown increased dPAG activity in panic patients [53] or healthy volunteers exposed to a proximal threatening stimulus such as predator exposure [54]. Panic-like responses can be modulated by several neurotransmitters including serotonin, GABA, glutamate and nitric oxide [55]. Several studies have indicated the presence of TRPV1 in the PAG (see 'Acute Pain'), which may influence the panic response. Local injection of the TRPV1 antagonists capsazepine or SB366791 into the dPAG attenuated panic-like behaviour induced by electrical stimulation [47]. TRPV1 antagonism in the dlPAG had similar effects in three other animal models of panic induced by (1) local injection of the excitatory amino acid N-methyl-D-aspartate (NMDA), (2) local injection of the nitric oxide donor SIN-1 and (3) exposure to the open arms of the elevated T maze [56–58]. These results suggest that TRPV1 in the dlPAG facilitates defensive responses in threatening situations.

Obsessive Compulsive Disorder. Marble-burying behaviour (MBB) is a commonly used model for assessing compulsive activity in rodents [59]. Studies indicate that the burying behaviour in rodents is an unconditioned, species-specific defensive reaction which is not associated with physical danger, and to which animals do not habituate upon repeated testing. Thus, MBB models some of the clinical symptoms of obsessive compulsive disorder which is characterized by recurrent obsessions or compulsions that severely impair daily routine.

A recent study by Umathe et al. [60] investigated the effects of capsaicin and capsazepine, administered intracerebroventricularly, on MBB. This study revealed that capsaicin produced compulsive effects (increased marble buying), similar to those of high-dose anandamide, whereas capsazepine dose-dependently decreased the bury-

Table 2. Effects of pharmacological modulation of TRPV1 in animal models of anxiety and depression

Drug and dose	Route of Admin./target region in CNS	Test/effect on behaviour	Animal	Reference
Anxiety				
Olvanil 0.2–5.0 mg/kg	i.p.	EPM/anxiogenic	Rat	[40]
Capsazepine (1–10 μg/kg)	i.p.	EPM/anxiolytic	Rat	[40]
SB366791 (0.1–2.5 mg/kg); Olvanil (0.1 mg/kg); AA-5-HT (0.1–5 mg/kg)	i.p.	EPM/anxiolytic	Mouse	[41]
Capsazepine (1, 10, 30, 60 nmol)	i.c. (mPFC)	EPM and VCT/anxiolytic	Rat	[44]
Methanamide 0.1–10 μg Capsaicin 1–10 μg and 1 nmol	i.c (mPFC)	EPM/anxiogenic	Rat	[45]
AA-5-HT (0.25–0.5 nmol)	i.c. (BLA)	EPM/anxiolytic	Rat	[48]
AMG 9810 (0.003, 0.03, 0.3 μg)	i.c. (dHPC)	EPM/anxiolytic	Rat	[49]
Capsaicin (0.003, 0.03, 0.3 μg)	i.c. (dHPC)	EPM/anxiogenic	Rat	[49]
Capsazepine (0.2–2 nmol)	i.c. (vHPC)	EPM/anxiolytic	Rat	[46]
Capsaicin (0.01, 0.1, 1 nmol)	i.c. (dPAG)	EPM/anxiogenic	Mouse	[50]
Capsaicin (0.01, 0.1, 1 nmol)	i.c. (dlPAG)	EPM and VCT/anxiolytic	Rat	[47]
Capsazepine (1, 10, 30, 60 nmol) and 6-iodonordihydrocapsaicin (3 nmol)	i.c. (mPFC)	Conditioned fear/decrease in fear-related behaviour	Rat	[43]
Capsaicin 1–10 μg and 1 nmol	i.c (mPFC)	Conditioned fear/Increase in fear-related behaviour	Rat	[43]
Capsazepine (1–60 nmol) 30 nmol*	i.c. (dlPAG)	ETM/panicolytic-like effects	Rat	[56]
SB366791 (10 nmol) Capsazepine (0.1, 1, 10 nmol)	i.c. (dlPAG)	Escape threshold determination/panicolytic-like effects	Rat	[58]
Capsazepine (30 nmol)	i.c. (dlPAG)	Escape threshold determination/panicolytic-like effects	Rat	[57]
Capsazepine (100 μg)	i.c.v.	Abolished MBB	Mice	[60]
Depression				
Olvanil 0.2–5.0 mg/kg	i.p.	Antidepressant effect/Porsolt swimming test	Rat	[40]
Capsaicin (200 and 300 μg/mouse) and capsazepine (100 and 200 μg/mouse)	i.c.v.	Antidepressant effect/forced swim test	Mice	[63]
Capsaicin (0.1, 1, 2.5 mg/kg) and Olvanil (0.1, 1, 5 mg/kg)	i.p.	Antidepressant effect/forced swim test	Mice	[64]
RTX (0.25 μg/kg i.t.)	i.t.	Increased immobility (depressive-like)/forced swim test	Rat	[65]
Amitriptyline (10 mg/kg) and ketamine (50 mg/kg)	i.p.	Decreased the immobility caused by RTX (antidepressant effect)/forced swim test	Rat	[65]

i.p. = Intraperitoneal; i.t = intrathecal; i.c. = intracerebral; i.c.v = intracerebroventricular; ETM = elevated T maze; VCT = vogel conflict test; mPFC = medial prefrontal cortex; vHPC = ventral hippocampus; dHPC = dorsal hippocampus; BLA = basolateral amygdala; SB366791 = N-(3-methoxyphenyl)-4-chlorocinnamide.

Fig. 1. A synthesis of the literature reviewed herein on the role of TRPV1 in discrete brain regions in pain and anxiety related behaviour. Bold text indicates the activation of TRPV1 in that brain region and normal text indicates the blockade/desensitization of the receptor in that brain region. ↑ Denotes an increase in anxiety/pain related behaviour, and ↓ denotes a decrease in anxiety/pain related behaviour. * Denotes initial activation of the receptor followed by desensitisation. Light grey shading of a brain region on the brain section images denotes anxiety-related studies of TRPV1 in that region and black shading denotes pain-related studies of TRPV1 in that region. mPFC = Medial prefrontal cortex; vHPC = ventral hippocampus; dHPC = dorsal hippocampus; BLA = basolateral amygdala; PL-IL = prelimbic-infralimbic cortex.

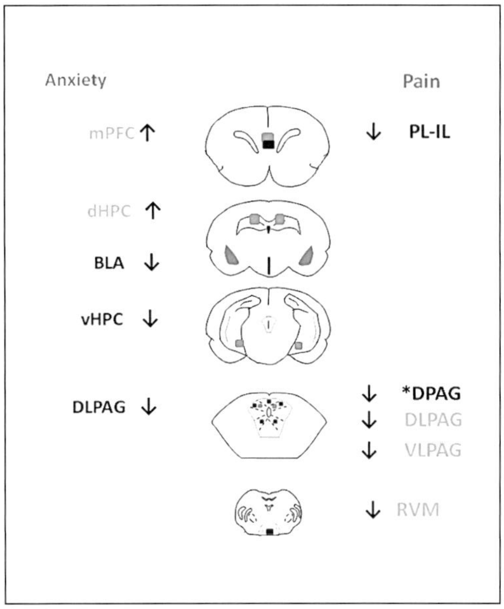

ing behaviour. These observations support the hypothesis that central TRPV1 might mediate the procompulsive effect of high doses of anandamide. Central administration of lower doses of anandamide, or drugs that elevate levels of anandamide (AM404/URB597), inhibited MBB, suggesting anticompulsive effects [60]. Pretreatment with a CB_1 receptor antagonist (intracerebroventricularly) abolished the anticompulsive effect of anandamide, whereas the TRPV1 antagonist capsazepine blocked the procompulsive effect of higher doses of anandamide [60]. Therefore, these results suggest that TRPV1 activation leads to an increase in obsessive compulsive disorder-like behaviour, with blockade of TRPV1 alleviating such behaviour. Further research employing additional animal models is required to determine the precise role of central TRPV1 in regulation of compulsive behaviour. Overall, preclinical studies show that TRPV1 plays a key role in anxiety-related behaviour. Given the high degree of comorbidity between anxiety disorders and chronic pain, and overlap in the TRPV1-expressing neuroanatomical substrates involved in both anxiety and pain, it is likely that TRPV1 also plays an important role in anxiety-pain interactions and further research in this area is warranted. Figure 1 represents a synthesis of the pain and anxiety literature reviewed above.

Transient Receptor Potential Subfamily V Member 1 and Depression

TRPV1 antagonists, administered systemically, have been shown to produce antidepressant-like effects in both rats and mice, suggesting a role for TRPV1 in depression [40, 61–64]. In the forced swim test, TRPV1 KO mice displayed lower immobility when compared to wild-type mice, indicating less behavioural despair in mice lacking TRPV1 [61]. Similarly, systemic administration of the TRPV1 antagonist capsazepine decreased the immobility time in a dose-dependent manner in the forced swim test [64]. Another paradigm for assessment of antidepressant-like activity is novelty-suppressed feeding. In this test, TRPV1 KO mice have decreased latency times when compared to wild-type mice, indicating an antidepressant phenotype [61]. Capsazepine is known to enhance antidepressant activity when administered to fluoxetine-treated mice at a subthreshold dose in the forced swim test [63]. Desensitization of supraspinal TRPV1 has also been shown to produce an antidepressant-like effect, as evidenced by the reduction in immobility time in the mouse forced swim test following intracerebroventricular injection of capsaicin at a dose that would have desensitized the receptor [63]. Further evidence for involvement of central TRPV1 comes from work demonstrating that intrathecal injections of a TRPV1-desensitizing dose of the agonist RTX also reduced immobility in the mouse forced swim test and inhibited the immobility induced by a lower dose of RTX [65]. Moreover, the antidepressants, amitriptyline and ketamine, administered intraperitoneally, inhibited the increase in forced swim test immobility (water at 41 °C) induced by a low dose of RTX administered subcutaneously [65]. These workers also demonstrated that water at 41 °C elicited less immobility than cooler water (26 °C), indicating that thermoregulatory sites do not contribute to immobility in the forced swim test. Finally, systemic administration of the TRPV1 agonist Olvanil reduced immobility in the rat forced swim test due to desensitisation [40].

Transient Receptor Potential Subfamily V Member 1 and Schizophrenia
Schizophrenia is a neurodevelopmental disorder and while there is currently a paucity of data directly linking TRPV1 to schizophrenia, there is evidence that TRPV1 plays a role in brain development. In this regard, potential links between TRPV1 and schizophrenia include dopaminergic mechanisms and cannabinoid mechanisms.

Effect of Capsaicin Treatment on Rat Brain Development. Deficits in pain sensation and altered vascular responsivity (flare response) to niacin have been reported in schizophrenic patients [66–69]. Primary afferent fibres are the subset of primary afferent neurons involved in both pain and flare responses that are sensitive to capsaicin treatment, i.e. TRPV1-containing afferents. Thus, patients with schizophrenia might have an abnormality in capsaicin-sensitive primary afferent neurons. A reduced neuropil (neuron density pertaining to brain grey matter) count, which is seen in schizophrenic patients, might also be due to reduced synaptic density in cortical regions arising from deficits in input from capsaicin-sensitive peripheral neurons [70, 71]. This hypothesis was recently tested in rats treated as neonates with capsaicin to destroy primary afferent neurons, which in turn would lead to an intrinsic somatosensory deprivation [72]. Within 5–7 weeks, locomotor activity in a novel environment was increased in rats that had been treated with capsaicin as neonates. Reduced brain weight was also observed, with a substantial decrease in the weight of the hippocampus and increase in neuronal density in cortical areas [72]. The changes reported were similar to those seen in schizophrenic subjects [72]. These findings suggest that neonatal capsaicin treatment may be useful for modelling aspects of schizophrenia.

Recently, Newson et al. [73] investigated the brain and behavioural responses in adult rats treated as neonates with capsaicin. The brain changes found at 5–7 weeks persisted in adult rats of 12 weeks, but diminished in older rats of 16–18 weeks. The rats exhibited increases in prepulse inhibition of acoustic startle at the age of 8 and 12 weeks. The study also reported that cutaneous plasma extravasation responses to niacin and prostaglandin D_2 were reduced in capsaicin-treated rats [72]. The neuroanatomical changes and reduced cutaneous plasma extravasation responses in capsaicin-treated rats resemble those observed in schizophrenic patients and suggest a potential role for TRPV1 in this psychiatric disorder.

Transient Receptor Potential Subfamily V Member 1 and Dopaminergic Mechanisms. Dopaminergic system hyperactivity is known to underlie the positive symptoms in schizophrenia [74]. As previously highlighted, TRPV1 has been identified in the cortex, hippocampus, basal ganglia, cerebellum, olfactory bulb, mesencephalon and hindbrain [9, 10]. In rat brain slices, activation of TRPV1 by capsaicin increases the rate of firing of dopamine neurons of the midbrain ventral tegmental area in a concentration-dependent manner [75]. Furthermore, in vivo experiments have shown that microinjection of capsaicin into the ventral tegmental area transiently increased dopamine release in the nucleus accumbens. Dopamine release by intraventral tegmental area administration of capsaicin was inhibited by the selective TRPV1 receptor antagonist, I-RTX, suggesting a role for mesencephalic TRPV1 in dopaminergic transmission [75]. Regulation of dopaminergic signalling in the brain's reward circuitry implies that TRPV1 could represent an important target for schizophrenia, affective disorders and addiction.

Transient Receptor Potential Subfamily V Member 1 and Cannabinoid Mechanisms. A link between cannabis use and early onset of the first episode of psychosis has been proposed, but the mechanisms that lead to psychosis are still unknown [76–78]. Systemic administration of the selective endocannabinoid reuptake inhibitors AM404 and VDM11, or the FAAH and TRPV1 inhibitor AA-5-HT, attenuated spontaneous hyperlocomotion in dopamine transporter KO mice. These hypolocomotor effects were significantly attenuated by coadministration of the TRPV1 antagonist capsazepine [79], highlighting an important role for TRPV1 in these responses. Recently, cannabidiol, a nonpsychotropic plant cannabinoid known to desensitize TRPV1 in vitro in epileptiform activity [80], may have therapeutic potential in psychosis, but the mechanisms underlying that effect are not clear and recent studies hypothesize that TRPV1 might be involved alongside CB_1 receptors [81].

Conclusion

The evidence for a role of brain TRPV1 in pain and neuropsychiatric disorders has been reviewed and discussed herein. Study of the role of TRPV1 in neuropsychiatric disorders is still at an early stage. There is a clear overlap of brain regions dually implicated in

TRPV1-mediated modulation of both neuropsychiatric disorders and pain. As such, alterations in the function of TRPV1 may underlie the pathophysiology of psychiatric and chronic pain conditions and their comorbidity. However, no studies to date have investigated the role of TRPV1 in this comorbidity and it is an obvious area for future research. In addition, further studies (preclinical and clinical) are needed to elucidate the precise neurobiological mechanisms through which supraspinal TRPV1 modulates pain and affect. Progress in these areas may pave the way towards TRPV1-targeting therapeutics for the treatment of pain and psychiatric disorders, and their comorbidity.

Acknowledgements

This work was funded by grants from Science Foundation Ireland (10/IN.1/B2976) and a PhD scholarship from the College of Science, National University of Ireland Galway.

References

1 Caterina MJ, Leffler A, Malmberg AB, Martin WJ, Trafton J, Petersen-Zeitz KR, Koltzenburg M, Basbaum AI, Julius D: Impaired nociception and pain sensation in mice lacking the capsaicin receptor. Science 2000;288:306–313.

2 Julius D: TRP channels and pain. Annu Rev Cell Dev Biol 2013;29:355–384.

3 Szallasi A, Sheta M: Targeting TRPV1 for pain relief: limits, losers and laurels. Expert Opin Investig Drugs 2012;21:1351–1369.

4 Ho KW, Ward NJ, Calkins DJ: TRPV1: a stress response protein in the central nervous system. Am J Neurodegener Dis 2012;1:1–14.

5 Lishko PV, Procko E, Jin X, Phelps CB, Gaudet R: The ankyrin repeats of TRPV1 bind multiple ligands and modulate channel sensitivity. Neuron 2007;54:905–918.

6 Garcia-Sanz N, Fernandez-Carvajal A, Morenilla-Palao C, Planells-Cases R, Fajardo-Sanchez E, Fernandez-Ballester G, Ferrer-Montiel A: Identification of a tetramerization domain in the C terminus of the vanilloid receptor. J Neurosci 2004;24:5307–5314.

7 Jung J, Hwang SW, Kwak J, Lee SY, Kang CJ, Kim WB, Kim D, Oh U: Capsaicin binds to the intracellular domain of the capsaicin-activated ion channel. J Neurosci 1999;19:529–538.

8 Price TJ, Louria MD, Candelario-Soto D, Dussor GO, Jeske NA, Patwardhan AM, Diogenes A, Trott AA, Hargreaves KM, Flores CM: Treatment of trigeminal ganglion neurons in vitro with NGF, GDNF or BDNF: effects on neuronal survival, neurochemical properties and TRPV1-mediated neuropeptide secretion. BMC Neurosci 2005;6:4.

9 Mezey E, Toth ZE, Cortright DN, Arzubi MK, Krause JE, Elde R, Guo A, Blumberg PM, Szallasi A: Distribution of mRNA for vanilloid receptor subtype 1 (VR1), and the VR1-like immunoreactivity in the central nervous system of the rat and human. Proc Natl Acad Sci U S A 2000;3655–3660.

10 Szabo T, Biro T, Gonzalez AF, Palkovits M, Blumberg PM: Pharmacological characterization of vanilloid receptor located in the brain. Brain Res Mol Brain Res 2002;98:51–57.

11 Sanchez JF, Krause JE, Cortright DN: The distribution and regulation of vanilloid receptor VR1 and VR1 5′ splice variant RNA expression in rat. Neuroscience 2001;107:373–381.

12 Cavanaugh DJ, Cavanaugh DJ, Chesler AT, Jackson AC, Sigal YM, Yamanaka H, Grant R, O'Donnell D, Nicoll RA, Shah NM, Julius D, Basbaum AI: Trpv1 reporter mice reveal highly restricted brain distribution and functional expression in arteriolar smooth muscle cells. J Neurosci 2011;31:5067–5077.

13 Han L, Ma C, Liu, Q, Weng HJ, Cui Y, Tang Z, Kim, Y, Nie H, Qu L, Patel KN, Li Z, McNeil B, He S, Guan Y, Xiao B, Lamotte RH, Dong X: A subpopulation of nociceptors specifically linked to itch. Nat Neurosci 2013;16:174–182.

14 Benninger F, Freund TF, Hajos N: Control of excitatory synaptic transmission by capsaicin is unaltered in TRPV1 vanilloid receptor knockout mice. Neurochem Int 2008;52:89–94.

15 Szallasi A: Autoradiographic visualization and pharmacological characterization of vanilloid (capsaicin) receptors in several species, including man. Acta Physiol Scand Suppl 1995;629:1–68.

16 Tominaga M, Caterina MJ, Malmberg AB, Rosen TA, Gilbert H, Skinner K, Raumann BE, Basbaum AI, Julius D: The cloned capsaicin receptor integrates multiple pain-producing stimuli. Neuron 1998;21:531–543.

17 Sudbury JR, Ciura S, Sharif-Naeini R, Bourque CW: Osmotic and thermal control of magnocellular neurosecretory neurons – role of an N-terminal variant of TRPV1. Eur J Neurosci 2010;32:2022–2030.

18 Goswami C, Rademacher N, Smalla KH, Kalscheuer V, Ropers HH, Gundelfinger ED, Hucho T: TRPV1 acts as a synaptic protein and regulates vesicle recycling. J Cell Sci 2010;123:2045–2057.

19 O'Sullivan GJ, O'Tuathaigh CM, Clifford JJ, O'Meara GF, Croke DT, Waddington JL: Potential and limitations of genetic manipulation in animals. Drug Discov Today Technol 2006;3:173–180.

20 Han P, Korepanova AV, Vos MH, Moreland RB, Chiu ML, Faltyneck CR: Quantification of TRPV1 protein levels in rat tissues to understand its physiological roles. J Mol Neurosci 2013;15:23–32.

21 Bajic D, Proudfit HK: Projections of neurons in the periaqueductal gray to pontine and medullary catecholamine cell groups involved in the modulation of nociception. J Comp Neurol 1999;405:359–379.

22 Tóth A, Boczan J, Kedei N, Lizanecz E, Bagi Z, Papp Z, Edes I, Csiba L, Blumberg PM: Expression and distribution of vanilloid receptor 1 (TRPV1) in the adult rat brain. Brain Res Mol Brain Res 2005;135:162–168.

23 Cristino L, de Petrocellis L, Pryce G, Baker D, Guglielmotti V, Di Marzo V: Immunohistochemical localization of cannabinoid type 1 and vanilloid transient receptor potential vanilloid type 1 receptors in the mouse brain. Neuroscience 2006;139:1405–1415.

24 Roberts JC, Davis JB, Benham CD: [3H]Resiniferatoxin autoradiography in the CNS of wild-type and TRPV1 null mice defines TRPV1 (VR-1) protein distribution. Brain Res 2004;195:176–183.

25 Fields HL: State-dependent opioid control of pain. Nat Rev Neurosci 2004;5:565–575.

26 Fields HL, Bry J, Hentall I, Zorman G: The activity of neurons in the rostral medulla of the rat during withdrawal from noxious heat. J Neurosci 1983;3:2545–2552.

27 Fields HL, Heinricher MM, Mason P: Neurotransmitters in nociceptive modulatory circuits. Annu Rev Neurosci 1991;14:219–245.

28 Palazzo E, de Novellis V, Marabese I, Cuomo D, Rossi F, Berrino L, Rossi F, Maione S: Interaction between vanilloid and glutamate receptors in the central modulation of nociception. Eur J Pharmacol 2002;439:69–75.

29 Starowicz K, Maione S, Cristino L, Palazzo E, Marabese I, Rossi F, de Novellis V, and Di Marzo V: Tonic endovanilloid facilitation of glutamate release in brainstem descending antinociceptive pathways. J Neurosci 2007;27:13739–13749.

30 Maione S, Bisogno T, de Novellis V, Palazzo E, Cristino L, Valenti M, Petrosino S, Guglielmotti V, Rossi F, Di Marzo V: Elevation of endocannabinoid levels in the ventrolateral periaqueductal grey through inhibition of fatty acid amide hydrolase affects descending nociceptive pathways via both cannabinoid receptor type 1 and transient receptor potential vanilloid type-1 receptors. J Pharmacol Exp Ther 2006;316:969–982.

31 McGaraughty S, Chu KL, Bitner RS, Martino B, El Kouhen R, Han P, Nikkel AL, Burgard EC, Faltynek CR, Jarvis MF: Capsaicin infused into the PAG affects rat tail flick responses to noxious heat and alters neuronal firing in the RVM. J Neurophysiol 2003;90: 2702–2710.

32 Liao HT, Lee HJ, Ho YC, Chiou LC: Capsaicin in the periaqueductal gray induces analgesia via metabotropic glutamate receptor-mediated endocannabinoid retrograde disinhibition. Br J Pharmacol 2011; 163:330–345.

33 de Novellis V, Palazzo E, Rossi F, De Petrocellis L, Petrosino S, Guida F, Luongo L, Migliozzi A, Cristino L, Marabese I, Starowicz K, Di Marzo V, Maione S: The analgesic effect of N-arachidonoyl-serotonin, a FAAH inhibitor and TRPV1 receptor antagonist, associated with changes in rostral ventromedial medulla and locus coeruleus cell activity in rats. Neuropharmacology 2008;55:1105–1113.

34 de Novellis V, Luongo L, Guida F, Cristino L, Palazzo E, Russo R, Marabese I, D'Agostino G, Calignano A, Rossi F, Di Marzo V, Maione S: Effects of intraventrolateral periaqueductal grey palmitoylethanolamide on thermoceptive threshold and rostral ventromedial medulla cell activity. Eur J Pharmacol 2012;676:41–50.

35 Cui M, Honore P, Zhong C, Gauvin D, Mikusa J, Hernandez G, Chandran P, Gomtsyan A, Brown B, Bayburt EK, Marsh K, Bianchi B, McDonald H, Niforatos W, Neelands TR, Moreland RB, Decker MW, Lee CH, Sullivan JP, Faltynek CR: TRPV1 receptors in the CNS play a key role in broad-spectrum analgesia of TRPV1 antagonists. J Neurosci 2006;26: 9385–9393.

36 da Silva KA, Manjavachi MN, Paszcuk AF, Pivatto M, Viegas C Jr, Bolzani VS, Calixto JB: Plant derived alkaloid (–)-cassine induces anti-inflammatory and anti-hyperalgesics effects in both acute and chronic inflammatory and neuropathic pain models. Neuropharmacology 2012;62:967–977.

37 Giordano C, Cristino L, Luongo L, Siniscalco D, Petrosino S, Piscitelli F, Marabese I, Gatta L, Rossi F, Imperatore R, Palazzo E, de Novellis V, Di Marzo V, Maione S: TRPV1-dependent and -independent alterations in the limbic cortex of neuropathic mice: impact on glial caspases and pain perception. Cereb Cortex 2012;22:2495–2518.

38 de Novellis V, Vita D, Gatta L, Luongo L, Bellini G, De Chiaro M, Marabese I, Siniscalco D, Boccella S, Piscitelli F, Di Marzo V, Palazzo E, Rossi F, Maione S: The blockade of the transient receptor potential vanilloid type 1 and fatty acid amide hydrolase decreases symptoms and central sequelae in the medial prefrontal cortex of neuropathic rats. Mol Pain 2011; 7:7.
39 Hong S, Fan J, Kemmerer ES, Evans S, Li Y, Wiley JW: Reciprocal changes in vanilloid (TRPV1) and endocannabinoid (CB1) receptors contribute to visceral hyperalgesia in the water avoidance stressed rat. Gut 2009;58:202–210.
40 Kasckow JW, Mulchahey JJ, Geracioti TD Jr: Effects of the vanilloid agonist olvanil and antagonist capsazepine on rat behaviors. Prog Neuropsychopharmacol Biol Psychiatry 2004;28:291–295.
41 Micale V, Cristino L, Tamburella A, Petrosino S, Leggio GM, Drago F, Di Marzo V: Anxiolytic effects in mice of a dual blocker of fatty acid amide hydrolase and transient receptor potential vanilloid type-1 channels. Neuropsychopharmacology 2009;34:593–606.
42 Marsch R, Foeller E, Rammes G, Bunck M, Kossl M, Holsboer F, Zieglgans-Berger W, Landgraf R, Lutz B, Wotjak CT: Reduced anxiety, conditioned fear, and hippocampal long-term potentiation in transient receptor potential vanilloid type 1 receptor-deficient mice. J Neurosci 2007;27:832–839.
43 Terzian AL, dos Reis DG, Guimaraes FS, Correa FM, Resstel LB: Medial prefrontal cortex transient receptor potential vanilloid type 1 (TRPV1) in the expression of contextual fear conditioning in Wistar rats. Psychopharmacology 2014;231:149–157.
44 Aguiar DC, Terzian AL, Guimaraes FS, Moreira FA: Anxiolytic-like effects induced by blockade of transient receptor potential vanilloid type 1 (TRPV1) channels in the medial prefrontal cortex of rats. Psychopharmacology 2009;205:217–225.
45 Rubino T, Realini N, Castiglioni C, Guidali C, Vigano D, Marras E, Petrosino S, Perletti G, Maccarrone M, Di Marzo V, Parolaro D: Role in anxiety behavior of the endocannabinoid system in the prefrontal cortex. Cereb Cortex 2008;18:1292–1301.
46 Santos CJ, Stern CA, Bertoglio LJ: Attenuation of anxiety-related behaviour after the antagonism of transient receptor potential vanilloid type 1 channels in the rat ventral hippocampus. Behav Pharmacol 2008;19:357–360.
47 Terzian AL, Aguiar DC, Guimaraes FS, Moreira FA: Modulation of anxiety-like behaviour by transient receptor potential vanilloid type 1 (TRPV1) channels located in the dorsolateral periaqueductal gray. Eur Neuropsychopharmacol 2009;19:188–195.
48 John CS, Currie PJ: N-arachidonoyl-serotonin in the basolateral amygdala increases anxiolytic behavior in the elevated plus maze. Behav Brain Res 2012;233: 382–388.
49 Hakimizadeh E, Oryan S, Hajizadeh Moghaddam A, Shamsizadeh A, Roohbakhsh A: Endocannabinoid system and TRPV1 receptors in the dorsal hippocampus of the rats modulate anxiety-like behaviors. Iran J Basic Med Sci 2012;15:795–802.
50 Mascarenhas DC, Gomes KS, Nunes-de-Souza RL: Anxiogenic-like effect induced by TRPV1 receptor activation within the dorsal periaqueductal gray matter in mice. Behav Brain Res 2013;250:308–315.
51 Blanchard DC, Blanchard RJ: Defensive behaviors, fear and anxiety; in Blanchard RJ, Blanchard DC, Griebel G, Nutt D (eds): Handbook of Anxiety and Fear, ed 1. Amsterdam, Elsevier, 2008, pp 63–99.
52 Schenberg LC, Bittencourt AS, Sudre EC, Vargas LC: Modeling panic attacks. Neurosci Biobehav Rev 2001;25:647–659.
53 Del-Ben CM, Graeff FG: Panic disorder: is the PAG involved? Neural Plast 2009;2009:108135.
54 Mobbs D, Petrovic P, Marchant JL, Hassabis D, Weiskopf N, Seymour B, Dolan RJ, Frith CD: When fear is near: threat imminence elicits prefrontal periaqueductal gray shifts in humans. Science 2007;317: 1079–1083.
55 Moreira FA, Guimaraes FS: Benzodiazepine receptor and serotonin 2A receptor modulate the aversive-like effects of nitric oxide in the dorsolateral periaqueductal gray of rats. Psychopharmacology (Berl) 2004;176:362–368.
56 Almeida-Santos AF, Moreira FA, Guimaraes FS, Aguiar DC: Role of TRPV1 receptors on panic-like behaviors mediated by the dorsolateral periaqueductal gray in rats. Pharmacol Biochem Behav 2013;105: 166–172.
57 Lisboa SF, Guimaraes FS: Differential role of CB1 and TRPV1 receptors on anandamide modulation of defensive responses induced by nitric oxide in the dorsolateral periaqueductal gray. Neuropharmacology 2012;62:2455–2462.
58 Casarotto PC, Terzian AL, Aguiar DC, Zangrossi H, Guimaraes FS, Wotjak CT, Moreira FA: Opposing roles for cannabinoid receptor type-1 (CB(1)) and transient receptor potential vanilloid type-1 channel (TRPV1) on the modulation of panic-like responses in rats. Neuropsychopharmacology 2012;37:478–486.
59 Broekkamp C, Rijk HW, Joly-Gelouin D, Lloyd KL: Major tranquilizers can be distinguished from minor tranquillizers on the basis of effects on marble burying behavior and swim-induced grooming in mice. Eur J Pharmacol 1986;126:223–229.
60 Umathe SN, Manna SS, Jain NS: Endocannabinoid analogues exacerbate marble-burying behavior in mice via TRPV1 receptor. Neuropharmacology 2012; 62:2024–2033.

61 You IJ, Jung YH, Kim MJ, Kwon SH, Hong SI, Lee SY, Jang CG: Alterations in the emotional and memory behavioral phenotypes of transient receptor potential vanilloid type 1-deficient mice are mediated by changes in expression of 5-HT(1)A, GABA(A), and NMDA receptors. Neuropharmacology 2012; 62:1034–1043.
62 Kulisch C, Albrecht D: Effects of single swim stress on changes in TRPV1-mediated plasticity in the amygdala. Behav Brain Res 2013;236:344–349.
63 Manna SS, Umathe SN: A possible participation of transient receptor potential vanilloid type 1 channels in the antidepressant effect of fluoxetine. Eur J Pharmacol 2012;685:81–90.
64 Hayase T: Differential effects of TRPV1 receptor ligands against nicotine-induced depression-like behaviors. BMC Pharmacol 2011;11:6.
65 Abdelhamid RE, Kovács KJ, Nunez MG, Larson AA: Depressive behavior in the forced swim test can be induced by TRPV1 receptor activity and is dependent on NMDA receptors. Pharmacol Res 2014;79: 21–27.
66 Kudoh A, Ishihara H, Matsuki A: Current perception thresholds and postoperative pain in schizophrenic patients. Reg Anesth Pain Med 2000;25: 475–479.
67 Blumensohn R, Ringler D, Eli I: Pain perception in patients with schizophrenia. J Nerv Ment Dis 2002; 190:481–483.
68 Waldo MC: Co-distribution of sensory gating and impaired niacin flush response in the parents of schizophrenics. Schizophr Res 1999;40:49–53.
69 Messamore E, Hoffman WE, Janowsky A: The niacin skin flush abnormality in schizophrenia: a quantitative dose-response study. Schizophr Res 2003;62: 251–258.
70 Selemon LD, Rajkowska G, Goldman-Rakic PS: Elevated neuronal density in prefrontal area 46 in brains from schizophrenic patients: application of a 3-dimensional, stereologic counting method. J Comp Neurol 1998;392:402–412.
71 Selemon LD, Goldman-Rakic PS: The reduced neuropil hypothesis: a circuit based model of schizophrenia. Biol Psychiatry 1999;45:17–25.
72 Newson P, Lynch-Frame A, Roach R, Bennett S, Carr V, Chahl LA: Intrinsic sensory deprivation induced by neonatal capsaicin treatment induces changes in rat brain and behaviour of possible relevance to schizophrenia. Br J Pharmacol 2005;146:408–418.
73 Newson PN, van den Buuse M, Martin S, Lynch-Frame A, Chahl LA: Effects of neonatal treatment with the TRPV1 agonist, capsaicin, on adult rat brain and behaviour. Behav Brain Res 2014;272:55–65.
74 Carlsson A, Lindqvist M: Effect of chlorpromazine or haloperidol on formation of 3-methoxytyramine and normetanephrine in mouse brain. Acta Pharmacol Toxicol 1963;20:140–144.
75 Marinelli S, Pascucci T, Bernardi G, Puglisi-Allegra S, Mercuri NB: Activation of TRPV1 in the VTA excites dopaminergic neurons and increases chemical and noxious-induced dopamine release in the nucleus accumbens. Neuropsychopharmacology 2005;30: 864–870.
76 Weiser M, Noy S: Interpreting the association between cannabis use and increased risk for schizophrenia. Dialogues Clin Neurosci 2005;7:81–85.
77 Gambi F, De Berardis D, Sepede G, Quartesan R, Calcagni E, Salerno RM, Conti CM, Ferro FM: Cannabinoid receptors and their relationships with neuropsychiatric disorders. Int J Immunopathol Pharmacol 2005;18:15–19.
78 Schuckit MA: Comorbidity between substance use disorders and psychiatric conditions. Addiction 2006;101(suppl 1):76–88.
79 Tzavara ET, Li DL, Moutsimilli L, Bisogno T, Di Marzo V, Phebus LA, Nomikos GG, Giros B: Endocannabinoids activate transient receptor potential vanilloid 1 receptors to reduce hyperdopaminergia related hyperactivity: therapeutic implications. Biol Psychiatry 2006;59:508–515.
80 Iannotti FA, Hill CL, Leo A, Alhusaini A, Soubrane C, Mazzarella E, Russo E, Whalley BJ, Di Marzo V, Stephens GJ: nonpsychotropic plant cannabinoids, cannabidivarin (CBDV) and cannabidiol (CBD), activate and desensitize transient receptor potential vanilloid 1 (TRPV1) channels in vitro: potential for the treatment of neuronal hyperexcitability. ACS Chem Neurosci 2014;5:1131–1141.
81 Campos AC, Moreira FA, Gomes FV, Del Bel EA, Guimarães FS: Multiple mechanisms involved in the large-spectrum therapeutic potential of cannabidiol in psychiatric disorders. Philos Trans R Soc Lond B Biol Sci 2012;367:3364–3378.

Prof. David P. Finn
Pharmacology and Therapeutics, School of Medicine, National University of Ireland
University Road
Galway (Ireland)
E-Mail david.finn@nuigalway.ie

Finn DP, Leonard BE (eds): Pain in Psychiatric Disorders. Mod Trends Pharmacopsychiatry.
Basel, Karger, 2015, vol 30, pp 94–102 (DOI: 10.1159/000435935)

Involvement of Neuroactive Steroids in Pain, Depression and Anxiety

Katherine Mifflin[a, b] · Curtis Benson[a, b] · Bradley Kerr[a–c] · Feyza Aricioglu[d] · Mesut Cetin[e] · Serdar Dursun[a, c] · Glen Baker[a, c]

[a]Department of Anesthesiology and Pain Research, [b]Neuroscience and Mental Health Institute, and [c]Neurochemical Research Unit, Department of Psychiatry, University of Alberta, Edmonton, Alta., Canada; [d]Department of Pharmacology and Psychopharmacology Research Unit, Marmara University, Istanbul, and [e]Department of Psychiatry, GATA Haydarpasa Training Hospital, Istanbul, Turkey

Abstract

Comorbidity between major depressive disorder (MDD), anxiety (generalized anxiety, panic disorder, social anxiety disorder) and pain is a major complicating factor in the diagnosis and treatment of psychiatric and neurological disorders. Although numerous neurotransmitters and/or neuromodulators may be involved, abnormalities in the GABAergic and glutamatergic systems seem to be a common factor in all these disorders. Neuroactive steroids (NASs) have been the object of considerable interest in this area in recent years since they appear to act predominantly on GABA-A and glutamate NMDA receptors. An overview of the possible involvement of NASs in MDD, anxiety and pain is provided in this chapter.

There is a high degree of comorbidity between pain and psychiatric disorders such as depression and anxiety [1–6 and other chapters in this book], and although this relationship is well established, the mechanisms remain unclear. In recent years, there has been a great deal of interest in the role of neuroactive steroids (NASs) in the etiology and pharmacotherapy of not only depression and anxiety, but also in pain syndromes [7–21]. This chapter will focus on neuropathic pain, which arises as a consequence of a lesion or a disease affecting the somatosensory systems [4, 22, 23]. Such pain usually arises from peripheral nerve injury or a metabolic disease such as diabetes [4].

Although numerous neurotransmitters and neuromodulators have been proposed to be involved in the etiology of major depressive disorder (MDD), anxiety disorders (generalized anxiety disorder, panic disorder, social anxiety disorder) and neuropathic pain, it is interesting that the disorders also seem to have in common possible

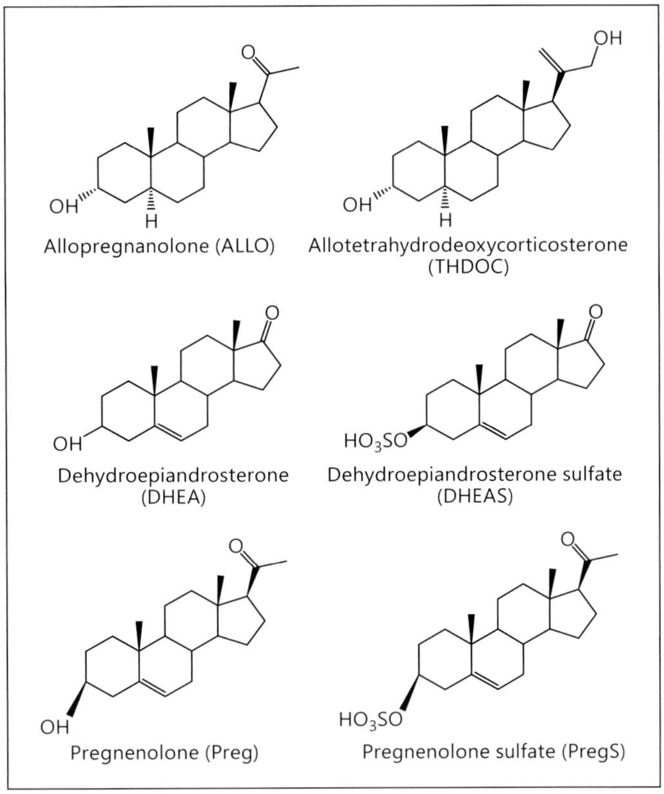

Fig. 1. Chemical structures of some NASs.

dysfunction of the GABAergic and/or glutamatergic systems. Hypofunction of the former system and hyperfunctioning of the latter have been proposed in all three disorders [11, 24–36]. Given the well-known action of the NASs on GABA-A and NMDA glutamate receptors [10, 37–42], the possible relationship between NASs and the comorbidity of MDD, anxiety and pain deserves further attention.

Neuroactive Steroids

NASs (fig. 1) are steroids that exert rapid, nongenomic effects through interactions with neurotransmitter receptors [9, 38, 41, 42]. NASs can originate as circulating steroid hormones or they can be synthesized in the brain from cholesterol [38] (fig. 2). These cholesterol-derived steroids in the brain are called neurosteroids. Hereafter in this chapter, the terms NAS and neurosteroid will be used interchangeably. A large body of literature exists demonstrating that several NASs act as allosteric modulators at various receptors, but most of the research has focused on the GABA-A receptor and the NMDA glutamate receptor [10, 19, 20, 38–43]. Some NASs act as positive and some as negative allosteric modulators at these receptors (fig. 3), and the balance in the levels of such positive and negative allosteric modulators is recognized as an

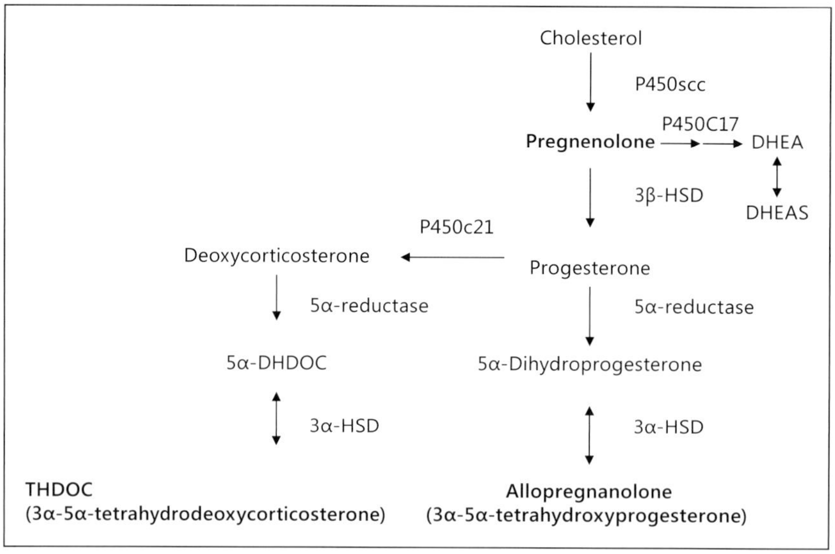

Fig. 2. Synthesis of some NASs.

NASs as allosteric modulators: examples

Allopregnanolone (ALLO), THDOC
Positive modulators at the GABA-A receptor (very potent effect, providing anxiolytic actions). Progesterone is also a positive modulator here

DHEA
Positive modulator at the NMDA glutamate receptor

DHEAS, pregnenolone sulfate
Negative modulators at the GABA-A receptor
Positive modulators at the NMDA glutamate receptor

Pregnenolone
Positive modulator at the NMDA receptor

Fig. 3. Examples of NASs as allosteric modulators.

important factor in the etiology and pharmacotherapy of several psychiatric disorders, and is postulated to be critical in chronic pain conditions [9, 10, 35–37, 40–43].

Neurosteroid synthesis begins with cholesterol's conversion into pregnenolone (fig. 2). Regulation of the formation of pregnenolone from cholesterol is mediated by the translocation of TSPO (formerly known as peripheral or mitochondrial benzodiazepine receptor) and is dependent on the enzyme P450scc [20]. Pregnenolone can then be converted to a variety of different steroids (fig. 2). The enzyme 3β-HSD converts pregnenolone to progesterone (PROG), which can then be subsequently reduced by the enzyme 5α-reductase, leading to the formation of 5α-dihydroprogesterone. The enzyme 3α-HSD can then act on 5α-dihydroprogesterone, converting it to allopregnanolone (ALLO; fig. 2).

Allosteric Modulation of GABA-A and NMDA Receptors

The delicate balance between GABAergic and glutamatergic systems is crucial for normal central nervous system functions [44, 45]. ALLO is a very potent positive modulator of GABA-A receptors and has strong anxiolytic properties [7, 8, 46]. It also holds promise as an analgesic [20, 47, 48]. Other examples of neurosteroids that act as positive allosteric modulators of GABA-A receptors are PROG and THDOC (3α,5α-tetrahydrodeoxycorticosterone; fig. 3). Evidence suggests that GABA-A receptors that contain the δ-subunit are particularly sensitive to the positive modulatory effects of these neurosteroids [19, 41]. Once converted from cholesterol, pregnenolone may be converted to pregnenolone sulfate or, via the enzyme P450C17, into dehydroepiandrosterone (DHEA) or its sulfated form, DHEAS (fig. 2). DHEA, DHEAS and pregnenolone sulfate act as positive allosteric modulators of NMDA receptors (fig. 3). DHEAS and pregnenolone sulfate also have the capacity to exert negative allosteric modulation of GABA-A receptors, adding to their capacity to promote excitability in the central nervous system. It should also be noted that pregnenolone itself can also act as a positive modulator of the NMDA receptor.

Neuroactive Steroids, Depression, Anxiety and Pain

Animal studies have demonstrated that neuropathic pain-induced mood disorders can be attenuated by anxiolytics such as etizolam or antidepressants such as imipramine or the selective serotonin reuptake inhibitor paroxetine [49, 50]. All three of these drugs have also been reported to have effects on GABA-A receptor subunit isoforms or brain levels of GABA [51–53]. Interestingly, the MAO-inhibiting antidepressant phenelzine, which is also effective in treating panic disorder and social anxiety disorder, also causes a marked increase in brain levels of GABA [54] and reduces symptoms of pain in an animal model of multiple sclerosis [Benson et al., unpublished]. Antidepressants of various types are often used in the treatment of pain [49, 55, 56]. ALLO is the most extensively studied NAS with regard to MDD, anxiety and pain syndromes. It is a potent positive GABA-A receptor modulator with strong anxiolytic properties [19, 39, 48]. Reduced levels of ALLO in plasma or cerebrospinal fluid have been reported to be associated with MDD and anxiety disorders [39, 48, 57, 58], and Schule et al. [19] have proposed that its regulation of emotion and its potential therapeutic use in depression and anxiety may not only involve GABAergic mechanisms, but also enhancement of neurogenesis, myelination and neuroprotection and effects on the hypothalamic-pituitary-adrenal axis. Brain and body fluid levels of ALLO and THDOC are increased by several antidepressants [18, 57, 58], and their precursor PROG appears to exert protective effects in peripheral neuropathy [59]. Both ALLO and PROG exert protective effects in spinal cord trauma, and ALLO appears to produce these effects through a mechanism involving GABA-A receptors [48]. Protective effects of ALLO have been reported in animal models of peripheral diabetic neuropathy [11, 47, 48]. It is interesting that ALLO is able to counteract myelin abnormalities observed in peripheral nerves of rats [60], particularly since there is currently a great deal of interest in possible myelination problems in MDD. Certainly depressive symptoms are common in demyelinating disorders like multiple sclerosis; interestingly, decreased synthesis of ALLO, THDOC and DHEA has been reported in multiple sclerosis [61].

ALLO reduces neuropathic symptoms such as allodynia and hyperalgesia produced by antineoplastic drugs [8, 17] or spinal nerve ligation [7], and Patte-Mensah et al. [62] have reported analgesic, neuroprotective, antidepressant and anxiolytic effects of ALLO. Oxytocin is proposed to produce spinal analgesia by stimulating ALLO synthesis [63].

In a study of spinal cord GABA-A receptors in an inflammatory pain situation, Poisbeau et al. [12] showed that such peripheral inflammation activates production of 5-α-reduced neurosteroids, which leads to a prolongation of GABA-A receptor-mediated synaptic currents and limitation of thermal hyperalgesia, suggesting that peripheral inflammation activates an endogenous NAS-based antinociceptive control. In related studies, Patte-Mensah et al. [13] have shown that the P450scc gene responsible for neurosteroid biosynthesis is overexpressed during neuropathic pain provoked by sciatic nerve ligature, also suggesting modulation of chronic pain by endogenous neurosteroids.

Pathirathna et al. [14, 15] provided evidence from studies on rats that both T-type calcium channels and GABA-A channels are responsible for the potent peripheral analgesic effects of 5-α-reduced NASs and reported that these NASs alleviate thermal mechanical hyperalgesia in rats with neuropathic pain. In a study in which rats were subjected to a sciatic nerve single ligature constriction and treated with daily injections of PROG or vehicle, the PROG-treated rats did not develop mechanical allodynia and had a lower number of painful responses to cold stimulation as well as a reduced number of NRI NMDA receptor subunits (both phosphorylated and nonphosphorylated forms) [37]. Using a model of diabetic neuropathy (streptozotocin-treated rats), Leonelli et al. [11] found that PROG and dihydro-PROG provided protective effects on peripheral nerves. Charlet et al. [59] reported that PROG-derived neurosteroids elevated nociceptive thresholds in naïve animals and reduced mechanical allodynia and thermal heat hyperalgesia in the carrageenan model of inflammatory pain, an effect thought to be mediated by GABA-A receptors.

Strous et al. [64] reported the effectiveness of DHEA augmentation in treating depressive and anxiety symptoms of schizophrenia. Chronic treatment of rats with DHEA increases basal nociceptive thresholds, and Patte-Mensah et al. [65] suggested that androgenic metabolites of DHEA exerted analgesic effects while DHEA caused a rapid pronociceptive effect. Kibaly et al. [66] found decreased DHEA synthesis from pregnenolone in neuropathic spinal cord slices and, based on a series of experiments, proposed that DHEA synthesized in the cord controls pain mechanisms. George et al.

[67] reported that cerebrospinal fluid levels of pregnenolone are decreased in affective illness, especially during episodes of active depression. Severe pain has been reported to cause elevated serum levels of pregnenolone, but to result in a drop in such levels if the pain does not abate [68].

Mitchell et al. [16], in a study on lamina II neurons of the spinal cord, suggested that extrasynaptic GABA-A receptors may be a molecular target for the analgesic effects of neurosteroids. Mensah-Nyagan et al. [17] reviewed the evidence showing that the spinal cord is an active production center for NASs and proposed that regulation of neurosteroidogenesis in the spinal cord might be an adaptive mechanism in coping with chronic pain and/or neurodegenerative states. Substance P, a major mediator of painful signals, inhibits conversion of PROG to neuroactive metabolites in the spinal cord, thus interfering with GABA-A receptor activity in the modulation of nociceptive transmission [69]. Nuclear receptors of the liver X receptor family regulate adrenal steroidogenesis via their ability to control cholesterol homeostasis, and activation of these receptors with a synthetic ligand leads to increased levels of NASs and protection of the sciatic nerve from diabetes-induced neuropathy [70].

Conclusion

MDD, anxiety and pain are often comorbid, and an increased understanding of the reasons for these co-occurrences may facilitate diagnosis and treatment of patients with such comorbidities. NASs are allosteric modulators of GABA-A and/or NMDA receptors, and there is strong evidence supporting the involvement of these receptors in the etiology and/or pharmacotherapy of MDD, anxiety disorders such as generalized anxiety disorder, panic disorder and social anxiety disorder, as well as neuropathic pain; a greater knowledge of those interactions should result in more effective treatment of patients. Potential novel therapies for comorbid disorders could involve the use of (alone or in combination with currently used drugs) analogues of NASs or drugs that affect enzymes involved in NAS synthesis or catabolism or TSPO ligands that increase neurosteroidogenesis. It will also be of interest to learn more about the effects of nonpharmacological treatments for mood and anxiety disorders and pain on levels and activity of NASs. Sex differences with regard to NASs in pain syndromes are also worthy of further investigation since prevalence rates of most persistent pain conditions are higher in females than in males [71], and women report greater pain sensitivity to various noxious stimuli [71].

Acknowledgements

The authors are grateful to the CIHR of Canada, the Multiple Sclerosis Society of Canada and the University of Alberta for funding, and to Karyn Crawford for her expert secretarial assistance.

References

1 Berna C, Leknes S, Holmes EA, Edwards RR, Goodwin GM, Tracey I: Induction of depressed mood disrupts emotion regulation neurocircuitry and enhances pain unpleasantness. Biol Psychiatry 2010;67:1083–1090.

2 Subramaniam M, Vaingakar JA, Abdin E, Chong SA: Psychiatric morbidity in pain conditions: results from the Singapore Mental Health Study. Pain Res Manag 2013;18:185–190.

3 Means-Christensen AJ, Roy-Byrne PP, Sherbourne CD, Craske MG, Stein MB: Relationships among pain, anxiety, and depression in primary care. Depress Anxiety 2008;25:593–600.

4 Yalcin I, Barthas F, Barrot M: Emotional consequences of neuropathic pain: insight from preclinical studies. Neurosci Biobehav Rev 2014;47:154–164.

5 Zhang S, Jin X, You Z, Wang S, Lim G, Yang J, McCabe M, Lin N, Marota J, Chen L, Mao J: Persistent nociception induces anxiety-like behavior in rodents: role of endogenous neuropeptide S. Pain 2014;155:1504–1515.

6 Woo AKM: Depression and anxiety in pain. Rev Pain 2010;4:8–12.

7 Kawano T, SogaT, Chi H, Eguchi S, Yamakazi F, Kumagai N, Yokoyama M: Role of the neurosteroid allopregnanolone in the hyperalgesic behavior induced by painful nerve injury in rats. J Anesth 2011;25:942–945.

8 Meyer L, Patte-Mensah C, Taleb O, Mensah-Nyagan AG: Allopregnanolone prevents and suppresses oxaliplatin-evoked painful neuropathy: multi-parametric assessment and direct evidence. Pain 2011;152:170–181.

9 Baulieu EE: Neurosteroids: of the nervous system, by the nervous system, for the nervous system. Recent Prog Horm Res 1997;52:1–32.

10 Eser D, Romeo E, Baghai TC, di Michele F, Schule C, Pasini A, Zwanzger P, Padberg F, Rupprecht R: Neuroactive steroids as modulators of depression and anxiety. Neuroscience 2006;138:1041–1048.

11 Leonelli E, Bianchi R, Cavaletti G, et al: Progesterone and its derivatives are neuroprotective agents in experimental diabetic neuropathy: a multimodal analysis. Neuroscience 2007;144:1293–1304.

12 Poisbeau P, Patte-Mensah C, Keller AF, Barrot M, Breton J-D, Luis-Delgado OE, Freund-Mercier MJ, Mensah-Nyagan AG, Schlichter R: Inflammatory pain upregulates spinal inhibition via endogenous neurosteroid production. J Neurosci 2005;25:11768–11776.

13 Patte-Mensah C, Li S, Mensah-Nyagan AG: Impact of neurpathic pain on the gene expression and activity of cytochrome P450side-chain-cleavage in sensory neural networks. Cell Mol Life Sci 2004;61:2274–2284.

14 Pathirathna S, Brimelow BC, Jagodic MM, et al: New evidence that both T-type calcium channels and GABAA channels are responsible for the potent peripheral analgesic effects of 5alpha-reduced neuroactive steroids. Pain 2005;114:429–443.

15 Pathirathna S, Todorovic SM, Covey DF, et al: 5alpha-reduced neuroactive steroids alleviate thermal and mechanical hyperalgesia in rats with neuropathic pain. Pain 2005;117:326–339.

16 Mitchell EA, Gentet LJ, Dempster J, et al: GABAA and glycine receptor-mediated transmission in rat lamina II neurones: relevance to the analgesic actions of neuroactive steroids. J Physiol 2007;583:1021–1040.

17 Mensah-Nyagan AG, Kibaly C, Schaeffer V, et al: Endogenous steroid production in the spinal cord and potential involvement in neuropathic pain modulation. J Steroid Biochem Mol Biol 2008;109:286–293.

18 Eser D, Schule C, Baghai TC, Romeo E, Rupprecht R: Neuroactice steroids in depression and anxiety disorders: clinical studies. Neuroendocrinology 2006;84:244–254.

19 Schule C, Eser D, Baghai TC, Nothdurfter C, Kessler JS, Rupprecht R: Neuroactive steroids in affective disorders: target for novel antidepressant or anxiolytic drugs? Neuroscience 2011;191:55–77.

20 Jevtovic-Todorovic V, Covey DF, Todorovic SM: Are neuroactive steroids promising therapeutic agents in the management of acute and chronic pain? Psychoneuroendocrinology 2009;34(suppl 1):S178–S185.

21 Verdi J, Jafari-Sabet M, Mokhtari R, et al: The effect of progesterone on expression and development of neuropathic pain in a rat model of peripheral neuropathy. Eur J Pharmacol 2012;699:207–212.

22 Attal N, Cruccu G, Baron R, Haanpää M, Hansson P, Jensen TS, Nurmikko T; European Federation of Neurological Societies: EFNS guidelines on the pharmacological treatment of neuropathic pain: 2010 revision. Eur J Neurol 2010;17:1113–1188.

23 Woolf CJ, Mannion RJ: Neuropathic pain: aetiology, symptoms, mechanisms, and management. Lancet 1999;353:1959–1964.

24 Mohler H: The GABA system in anxiety and depression and its therapeutic potential. Neuropharmacology 2012;62:42–53.

25 Paul IA, Skolnick P: Glutamate and depression: clinical and preclinical studies. Ann N Y Acad Sci 2003;1003:250–272.

26 Mitchell ND, Baker GB: An update on the role of glutamate in the pathophysiology of depression. Acta Psychiatr Scand 2010;122:192–210.

27 Hashimoto K: The role of glutamate in the action of antidepressants. Prog Neuropsychopharmacol Biol Psychiatry 2011;35:1558–1568.

28 Amiel JM, Matthew SJ: Gluamate and anxiety disorders. Curr Psychiatry Rep 2007;9:278–283.
29 Cortese BM, Phan KL: The role of glutamate in anxiety related disorders. CNS Spectr 2005;10:820–830.
30 Gwak YS, Hulsebosch CE: GABA and central neuropathic pain following spinal cord injury. Neuropharmacology 2011;60:799–808.
31 Gegelashvili G, Bjerrum OJ: High-affinity glutamate transporters in chronic pain: an emerging therapeutic effect. J Neurochem 2014;131:712–730.
32 Larsson M, Broman J: Synaptic plasticity and pain: role of ionotropic glutamate receptors. Neuroscientist 2010;17:256–273.
33 Le Coz GM, Fiatte C, Anton F, Hanesch U: Differential neuropathic pain sensitivity and expression of spinal mediators in Lewis and Fischer 344 rats. BMC Neurosci 2014;15:35.
34 Sung B, Lim G, Mao J: Altered expression and uptake activity of spinal glutamate transporters after nerve injury contribute to the pathogenesis of neuropathic pain in rats. J Neurosci 2003;23:2899–2910.
35 Moore KA, Kohno T, Karchewski LA, Scholz J, Baba H, Woolf CJ: Partial peripheral nerve injury promotes a selective loss of GABAergic inhibition in the superficial dorsal horn of the spinal cord. J Neurosci 2002;22:6724–6731.
36 Zeilhofer HU, Möhler H, Di Lio A: GABAergic analgesia: new insights from mutant mice and subtype-selective agonists. Cell 2009;30:397–402.
37 Coronel MF, Labombarda F, Roig P, et al: Progesterone prevents nerve-induced allodynia and spinal NMDA receptor upregulation in rats. Pain Med 2011;12:1249–1261.
38 Mellon SH, Griffin LD, Compagnone NA: Biosynthesis and action of neurosteroids. Brain Res Rev 2001;37:3–12.
39 MacKenzie EM, Odontiadis J, Le Melledo J-M, Prior T, Baker GB: The relevance of neuroactive steroids in schizophrenia, depression, and anxiety disorders. Cell Mol Neurobiol 2007;27:541–574.
40 Le Mellédo, J-M, Baker GB: Neuroactive steroids and anxiety disorders. J Psychiatr Neurosci 2002;27:161–167.
41 Belelli D, Herd MB, Mitchell EA, Peden DR, Vardy AW, Gentet L, Lambert JJ: Neuroactive steroids and inhibitory neurotransmission: mechanisms of action and physiological relevance. Neuroscience 2006;138:821–829.
42 Dubrovsky BO: Steroids, neuroactive steroids and neurosteroids in psychopathology. Prog Neuropsychopharmacol Biol Psychiatry 2005;29:169–192.
43 Aouad M, Charlet A, Rodeau J-L, Poisbeau P: Reduction and prevention of vincristine-induced neuropathic pain symptoms by the non-benzodiazepine anxiolytic etifoxine are mediated by 3α-reduced neurosteroids. Pain 2009;147:54–59.
44 Kendell SF, Krystal JH, Sanacora G: GABA and glutamate systems as therapeutic targets in depression and mood disorders. Expert Opin Ther Targets 2005;9:153–168.
45 El-Ansary A, Al-Ayadhi L: GABAergic/glutamatergic imbalance relative to excessive neuroinflammation in autism spectrum disorders. J Neuroinflammation 2014;11:189–200.
46 Brot MD, Akwa Y, Purdy RH, Koob GF, Britton KT: The anxiolytic-like effects of the neurosteroid allopregnanolone. Eur J Pharmacol 1997;325:1–7.
47 Afrazi S, Esmaeili-Mahani S: Allopregnanolone suppresses diabetes-induced neuropathic pain and motor deficit through inhibition of GABAA receptor down-regulation in the spinal cord of diabetic rats. Iran J Basic Med Sci 2014;17:312–317.
48 Melcangi RC, Panzica GC: Allopregnanolone: state of the art. Prog Neurobiol 2014;113:1–5.
49 Matsuzawa-Yanagida K, Narita M, Nakajima M, Kuzumaki N, Niikura K, Nozaki H, Takagi T, Tamai E, Hareyama N, Terada M, Yamazaki M, Suzuki T: Usefulness of antidepressants for improving the neuropathic pain-like state and pain-induced anxiety through actions at different brain sites. Neuropsychopharmacology 2008;33:1952–1965.
50 Narita M: Chronic pain induces anxiety with concomitant changes in opoidergic function in the amygdala. Neuropsychopharmacology 2006;31:739–750.
51 Sanna E, Pau D, Tuveri F, Massa F, Maciocco E, Acquas C, Floris C, Fontana SN, Maira G, Biggio G: Molecular and neurochemical evaluation of the effects of etizolam on GABAA receptors under normal and stress conditions. Arzneimittelforschung 1999;49:88–95.
52 Tanay VM, Greenshaw AJ, Baker GB, Bateson AN: Common effects of chronically administered antipanic drugs on brainstem $GABA_A$ receptor subunit gene expression. Mol Psychiatry 2001;6:404–412.
53 Sanacora F, Mason GF, Rothman DL, Krystal JH: Increased occipital cortex GABA concentrations in depressed patients after therapy with selective serotonin reuptake inhibitors. Am J Psychiatry 2002;159:663–665.
54 Baker GB, Wong JT, Yeung JM, Coutts RT: Effects of the antidepressant phenelzine on brain levels of γ-aminobutyric acid (GABA). J Affect Disord 1991;21:207–211.
55 Dworkin RH, O'Connor AB, Backonja M, Farrar JT, Finnerup NB, Jensen TS, Kalso EA, Loeser JD, Miaskowski C, Nurmikko TJ, Portenoy RK, Rice AS, Stacey BR, Treede RD, Turk DC, Wallace MS: Pharmacologic management of neuropathic pain: evidence-based recommendations. Pain 2007;132:237–251.

56 Saarto T, Wiffen PJ: Antidepressants for neuropathic pain: a Cochrane review. J Neurol Neurosurg Psychiatry 2010;81:1372–1373.
57 Pinna G, Costa E, Guidotti A: Fluoxetine and norfluoxetine stereospecifically and selectively increase brain neurosteroid content at doses that are inactive on 5-HT reuptake. Psychopharmacology (Berl) 2006;186:362–372.
58 Romeo E, Strohle A, Spalletta G, di Michele F, Hermann B, Holsboer F, Pasini A, Rupprecht R: Effects of antidepressant treatment on neuroactive steroids in major depresssion. Am J Psychiatry 1998;155: 910–913.
59 Charlet A, Lasbennes F, Darbon P, Poisbeau P: Fast non-genomic effects of progesterone-derived neurosteroids on nociceptive thresholds and pain symptoms. Pain 2008;139:603–609.
60 Azcoitia I, Leonelli E, Magnaghi V, Veiga S, Garcia-Segura LM, Melcangi RC: Progesterone and its derivatives dihydroprogesterone and tetrahydroprogesterone reduce myelin fiber morphological abnormalities and myelin fiber loss in the sciatic nerve of aged rats. Neurobiol Aging 2003;24:853–860.
61 Noorbakhsh F, Ellestad KK, Maingat F, Warren KG, Han MH, Steinman L, Baker G, Power C: Impaired neurosteroid synthesis in multiple sclerosis. Brain 2011;134:2703–2721.
62 Patte-Mensah C, Meyer L, Taleb O, Mensah-Nyagan AG: Potential role of allopregnanolone for a safe and effective therapy of neuropathic pain. Prog Neurobiol 2013;113:70–78.
63 Juif PE, Breton J-D, Rajalu M, Charlet A, Goumon Y, Poisbeau P: Long-lasting spinal oxytocin analgesia is ensured by the stimulation of allopregnanalone synthesis which potentiates $GABA_A$ receptor-mediated synaptic inhibition. J Neurosci 2013;33:16617–16626.
64 Strous RD, Maayan R, Lapidus R, Stryjer R, Lustig M, Kotler MD, Weizman A: Dehydroepiandrosterone augmentation in the management of negative, depressive, and anxiety symptoms in schizophrenia. Arch Gen Psychiatry 2003;60:133–141.
65 Patte-Mensah C, Meyer L, Kibaly C, Mensah-Nyagan AG: Regulatory effect of dehydroepiandrosterone on spinal cord nociceptive function. Front Biosci 2010;2:1528–1537.
66 Kibaly C, Meyer L, Patte-Mensah C, Mensah-Nyagan AG: Biochemical and functional evidence for the control of pain mechanisms by dehydroepiandrosterone endogenously synthesized in the spinal cord. FASEB J 2008;22:93–104.
67 George MS, Guidotti A, Rubinow D, Pan B, Mikalauskas K, Post RM: CSF neuroactive steroids in affective disorders: pregnenolone, progesterone, and DBI. Biol Psychiatry 1994;35:775–780.
68 Tennant F: The physiologic effects of pain on the endocrine system. Pain Ther 2013;2:75–86.
69 Patte-Mensah C, Kibaly C, Mensah-Nyagan AG: Substance P inhibits progesterone conversion to neuroactive metabolites in spinal sensory circuit: a potential component of nociception. Proc Natl Acad Sci U S A 2005;102:9044–9049.
70 Cermenati G, Giatti S, Cavaletti G, Bianchi R, Maschi O, Persaresi M, Abbiati F, Volonterio A, Saez E, Caruso D, Melcangi RC, Mitro N: Activation of the liver X receptor increases neuroactive steroid levels and protects from diabetes-induced peripheral neuropathy. J Neurosci 2010;30:11896–11901.
71 Fillingham RB, King CD, Ribeiro-Dasilva MC, Rahim-Williams B, Riley JL 3rd: Sex, gender, and pain: a review of recent clinical and experimental findings. J Pain 2009;10:447–485.

Prof. Serdar Dursun
Neurochemical Research Unit
Department of Psychiatry, University of Alberta
Edmonton, AB T6G 2B7 (Canada)
E-Mail dursun@ualberta.ca

Finn DP, Leonard BE (eds): Pain in Psychiatric Disorders. Mod Trends Pharmacopsychiatry.
Basel, Karger, 2015, vol 30, pp 103–119 (DOI: 10.1159/000435936)

Visceral Pain and Psychiatric Disorders

Valeria D. Felice • Rachel D. Moloney • John F. Cryan •
Timothy G. Dinan • Siobhain M. O'Mahony

Department of Anatomy and Neuroscience, University College, Cork, Ireland

Abstract

The high comorbidity existing between visceral pain and psychiatric disorders such as depression and anxiety is well documented and it is gaining increasing interest among scientists. When visceral pain and psychiatric disorders are comorbid, they present a more debilitating condition than each disorder alone, impacting significantly on the quality of life of these patients. Despite several groups having shown that an overlapping pathophysiology exists between visceral pain and stress-related disorders the link between them is not clear yet. Moreover, it still remains to be elucidated if psychiatric conditions predispose the individual to develop visceral hypersensitivity or vice versa. The brain-gut-microbiome axis is the bidirectional communication between the CNS and the gastrointestinal tract. Alterations at different levels of this axis have been implicated in both visceral hypersensitivity and psychiatric disorders. Here we give an overview of what it is known about comorbid visceral pain and psychiatric disorders and provide evidence of potential overlapping pathophysiological mechanisms involved. Preclinical models of comorbid visceral pain and stress-related disorders are also discussed.

Visceral pain is the most common form of pain reported in the clinic [1]. Moreover, it is now recognised that visceral pain often coexists with psychiatric conditions such as depression and anxiety [2]. Patients suffering from visceral pain often present with emotional distress associated with alterations in the attentional, perceptional and affective responses [3]. The comorbidity between psychiatric disorders and visceral pain leads to a more debilitating condition than either alone, significantly impacting the quality of life and contributing to an economic burden. Although they share common psychological and biological factors, including similar neural pathways and neurotransmitters, the link between visceral pain and psychiatric disorders is not well understood [2]. Stressful events may either exacerbate pain, referred to as stress-

Table 1. The different manifestations of visceral pain and the prevalence of each type

Type	Disorder	Population prevalence, %	Reference
Abdominal pain	IBS	10–20	152
	Functional dyspepsia	20–40	153
	Functional abdominal pain syndrome	0.5–1.7	154
Chronic pelvic pain	Endometriosis	5–10	155
	Interstitial cystitis	2.7–6.5	156
	Chronic prostatitis	8.2	157
Chest pain	Musculoskeletal	20.4	158
	Reflux oesophagitis	13.4	158
	Cardiac syndrome X	30 (of angina patients)	159
	Costochondritis	13.1	158

induced hyperalgesia, or suppress it, referred to as stress-induced analgesia. This phenomenon depends on the duration, nature and intensity of the stressful stimulus [4–6].

Visceral Pain

The recent surge in interest by researchers and clinicians in visceral pain indicates a significant increase in the awareness and extent of visceral pain disorders. Visceral pain can range from the discomfort of indigestion to the excruciating pain of renal colic [7]. Much of our knowledge on pain processing has been obtained from studies of somatic pain rather than visceral, perhaps due to the difficulty in accessing visceral structures [7]. That there are comparatively fewer studies on visceral pain than somatic does not detract from its significance. There is considerable overlap in the processing pathways of both types of pain, but there are also important differences which contribute to differential treatment strategies [8]. Visceral pain results from the activation of nociceptors localised in thoracic and pelvic organs. Several stimuli can activate these receptors, including stretch, inflammation and oxygen starvation (ischaemia). This can be experimentally studied by stretching of the organ by inflating a balloon within the lumen [9]. Disorders of visceral pain significantly impact the quality of life and are associated with psychological disturbances as well as sleep and sexual dysfunction (table 1) [10].

Types of Visceral Pain

Visceral pain is associated with a wide variety of disorders including gallstones, acute pancreatitis, acute appendicitis, diverticulitis, painful functional bowel syndromes such as irritable bowel syndrome (IBS) and functional dyspepsia, inflammatory bowel disease, gastroesophageal reflux disease, interstitial cystitis/bladder pain syndrome, male chronic pelvic pain syndrome, and gynaecological pain (endometriosis, vulvodynia, menstrual pain, polycystic ovary syndrome) [11–14]. Moreover, and less commonly known, visceral pain also encompasses chronic chest pain and colic [15, 16]. Appreciating this wide array of disorders has allowed us to understand the complex nature of the pathophysiology of visceral pain, but has also induced considerable hurdles when aiming to fully understand the distinct subset of molecular mechanisms which underpin this phenomenon. For the remainder of this review we will specifically discuss gastrointestinal visceral pain associated with IBS. IBS is a functional gastrointestinal disorder that has significant visceral pain and psychiatric components and offers an opportunity to study the impact of pain on psychiatric disease and vice versa.

Stress and the Hypothalamic-Pituitary-Adrenal Axis

The hypothalamic-pituitary-adrenal (HPA) axis is activated in response to physiological and psychological challenges [17] and stimulates the release of glucocorticoids from the adrenal cortex [17]. Glucocorticoids interact with two receptors, the mineralocorticoid and glucocorticoid receptors, and release is mediated through the process of negative feedback: binding of these receptors at each level of the HPA axis as well as receptors within corticolimbic structures such as the medial prefrontal cortex and hippocampus facilitates termination of the HPA response [18]. Prolonged or excessive activation of the HPA axis creates strain, called 'allostatic load', on the body and brain [19]. Thus, an inability to adapt to stressful events can lead to pathological conditions. This is due not only to the qualities of the stressor, such as the duration, severity and the type of stressor, but also to factors that influence an organism's ability to adapt to the stressor. These include genetic factors, early-life experience, cognitive factors and environmental support [20]. Coping styles have been shown to influence the response of the neuroendocrine and neuroimmunological systems to stressful events [21], as well as hippocampus, medial prefrontal cortex, reward, and dorsal raphe nucleus circuits. Environmental manipulations (such as maternal separation), genetic factors, and sex differences have been shown to have long-lasting consequences on stress vulnerability and resilience [22].

Visceral Pain and Stress-Related Psychiatric Disorders

Stress-related psychiatric disorders including depression and anxiety are among the most disabling and prevalent medical conditions [23, 24]. These disorders are often comorbid with a variety of other systemic and somatic illnesses, ranging from cardiovascular disease [25] to alterations in gastrointestinal tract function [26, 150]. Despite the well-established association between stress and psychiatric disorders, the struggle to understand the complex processes by which stress mediates pathological changes, increasing vulnerability to disease, is on-going [27, 28]. Physical and psychological stressors have also been shown to increase the perception of visceral pain, which is referred to as visceral hypersensitivity [5, 6, 29, 30]. Psychological and pharmacological interventions that mediate stress perception have been demonstrated to ameliorate visceral pain symptoms [31, 32]. Interestingly, psychiatric disorders such as depression and anxiety often present comorbidly with increased visceral pain [33]. However, the reason and causality for this relationship is unclear; it is not known if stress-related disorders predispose to the development of visceral hypersensitivity or if altered emotional states arise in response to visceral hypersensitivity.

Impact of Gender

Gender differences in pain sensitivity and tolerance have been studied many times [34, 35]. Some clinical studies indicate that females have a lower tolerance to pain than men [35]. However, women have shown the ability to habituate to chronic exposure to painful stimuli [36]. Females are reported to experience a higher level of pain compared to men in different situations, such as spinal cord injury, in pressure pain applied to the neck [37], and visceral pain in IBS [38]. IBS is more common in females with a ratio of about 2:1 [39]. Stress-related disorders are also more prevalent in females [40, 41] with studies indicating a gender difference in stress-induced pain modulation [41]. Moreover, it was shown that the anti-inflammatory/analgesic compound, ibuprofen, had a reduced effect in young women compared to young men [42]. Gender differences in central activation patterns as well as cortical thickness are associated with pain [43, 44]. There is increased evidence in the literature that indicates males and females perceive pain differently when they were subjected to experimental or clinical pain [35]. Several factors have been suggested to explain the differences with the most common difference being the sex hormones [45, 46]. The fluctuating hormones affect women's perception of pain and are also associated with increased anxiety [47].

The Involvement of the Immune System in Visceral Pain and Psychiatric Disease

Pain sensation can be enhanced by inflammation, an effect due to the sensitisation of primary sensory nociceptive neurons. The most potent immunostimulatory cue produced by Gram-negative bacteria is lipopolysaccharide (LPS), an abundant outer wall glycolipid that is released locally and into circulation on bacterial lysis [48]. The stress signalling cascade utilises the presence of LPS, which acts via Toll-like receptor 4. This particular Toll-like receptor along with several others in the family are increased in the model of early-life stress, the maternal separation model [49]. This animal model displays comorbid visceral pain and depressive-like behaviours as well as a heightened systemic immune response [50]. Physiologically, the intestinal microbiota provides a source for endogenous LPS, i.e. during the stress response, and psychosocial stress challenge in mice enhances the gastrointestinal permeability and bacterial uptake from the gut [51]. Release of LPS from bacteria leads to an increase in gut permeability which positively correlates with visceral pain or stress-induced visceral hypersensitivity [52]. It is also well recognised that an association between psychiatric disorders and the immune system exists [53]. Chronic activation of the immune system can contribute to the development or exacerbation of depressive and anxiety symptoms [54]. Patients affected by depression present with increased levels of cytokines in plasma and cerebrospinal fluid [55, 56]. Moreover, increased levels of peripheral pro-inflammatory cytokines are observed in anxiety disorders [57]. Injection of LPS represents one of the approaches used to investigate the role of peripheral immune activation on mood disorders. Studies have shown that LPS-induced neuroinflammation is implicated in depressive-like symptoms in rodents [58], which are reversed following treatment with anti-inflammatory agents [58]. Moreover, LPS administration in healthy subjects has been shown to increase anxiety and negative mood [59].

The Brain-Gut-Microbiome Axis in Visceral Pain and Psychiatric Disease

The brain-gut-microbiome axis is the bidirectional communication between the central nervous system and the gastrointestinal tract which involves neural pathways, immune and endocrine mechanisms [60] (fig. 1). Under physiological conditions the brain-gut axis is responsible for the modulation of digestive processes (i.e. motility, secretion), immune function, perception and emotional response to visceral stimuli [61]. Perturbation of this axis results in alterations in the stress response and behaviour, and has been implicated in various disease states [33, 62–64]. The high comorbidity between stress-related psychiatric symptoms such as anxiety with gastrointestinal disorders including IBS and inflammatory bowel disease [62, 63] are further evidence of the impact of this axis. Recent studies have investigated the impact of gut microbiota on brain and behaviour. Interestingly, intestinal infection with the pathogenic bacterial *Campylobacter jejuni* induces activation of central brain areas such as the nucleus of the solitary tract and parabrachial nucleus, implicated in modulation of gastrointestinal sensory information [65]. Moreover, it has been shown that treatment with *Lactobacillus rhamnosus* results in alteration of GABA receptor expression in several central brain areas and a reduction in anxiety and depression-like behaviours [66]. The absence of the microbiota or its modification has been associated with alteration in HPA axis activity [67] as well as with anxiety-like behaviours. Germ-free mice display less anxiety-like behaviours than their conventionally colonised counterparts [68–70], while a normoanxious state can be reinstated with the introduction of a microbiota prior to critical time windows [70]. These mice also show an increased visceral sensitivity during colorectal distension which can be also be attenuated by

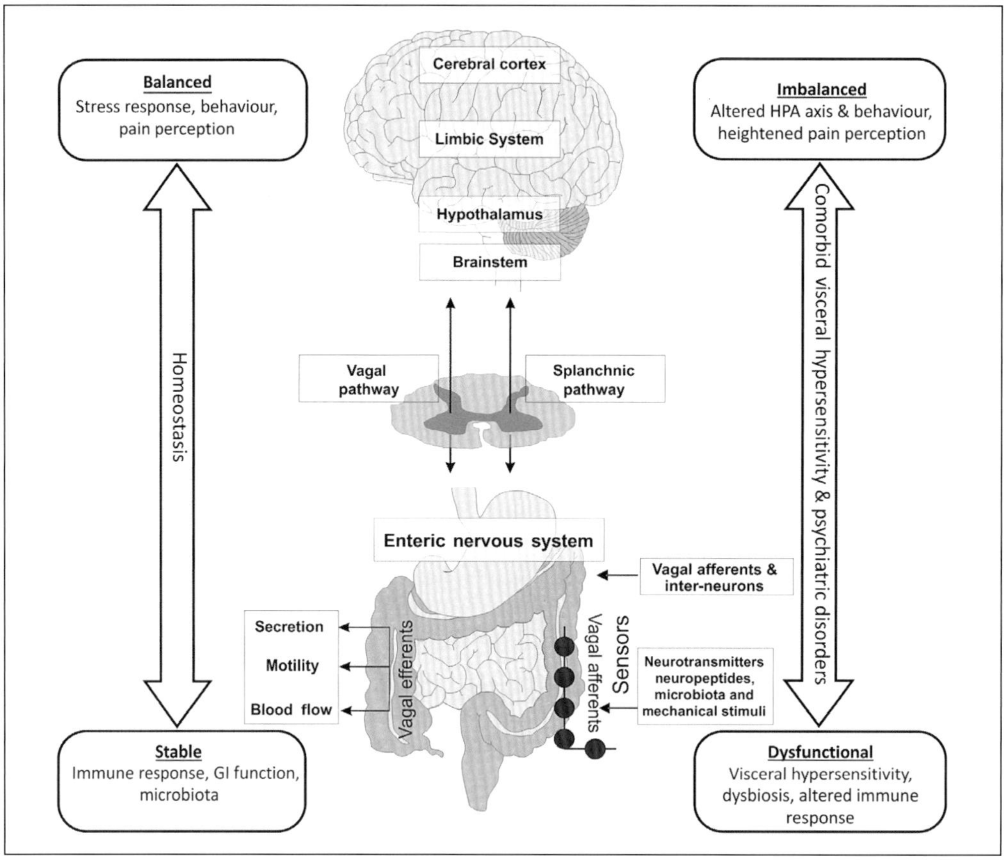

Fig. 1. The brain-gut-microbiota axis in health and disease. A bidirectional communication system between the brain and the gastrointestinal tract. The microbiota within the bowel is known to contribute to the development and functioning of this axis. This axis allows for events that occur in the central nervous system, such as psychiatric disorders, to impact on the gut and lead to dysfunction within the bowel such as an altered bacterial balance or dysbiosis, enhanced pain signalling and increased immune response. Furthermore, altered gastrointestinal function impacts on the brain which can lead to symptom perpetuation.

colonisation with bacteria from conventional mice. Destruction of the microbiota community in mice with an antimicrobial cocktail also reduces anxiety-like behaviours, while it has also been established that anxiety-like traits are transmissible following transfer of the microbiota between different mouse strains [71]. The association between the microbiota and depressive-like behaviours was shown preclinically following the administration of probiotic strains such as *L. rhamnosus* [66], *Bifidobacteria infantis* [72] and a formulation of *Lactobacillus helveticus* and *Bifidobacterium longum* [73]. Moreover, in healthy volunteers, the latter combination of probiotic strains alleviated psychological distress. Interestingly, further clinical evidence comes from the utility of antibacterial agents in the modulation of depression including minocycline [74, 75]. However, the mechanism of antidepressant action of minocycline could also be related to neuroprotection, suppression of microglial activation or anti-inflammatory actions [76].

The microbiota and manipulation of it has been associated with several types of pain including visceral pain [77–79], inflammatory pain [80], migraine [81] and autoimmune-related pain in rheumatoid arthritis [82]. Moreover, studies have shown that gastrointestinal infections and disease are associated with thoracic pain, headache [83] and migraine [81], indicating that the influence of the microbiota on pain may reach further than the gut. The majority of studies on pain and the microbiota have focused on gastrointestinal visceral pain, which is associated with an altered gut bacterial profile [84]. Lactic acid-producing bacteria such as bifidobacteria and lactobaccilus have been shown to reduce visceral pain in patients [85]. We have demonstrated that ablation of Gram-positive bacteria in the neonatal period is associated with the development of visceral pain in adult rats [78]. Verdu et al. [86] found similar results when they disrupted the microbiota in adult mice [86], and they also attenuated the visceral hypersensitivity with probiotic administration. Probiotic bacteria have shown efficacy in not only reducing visceral pain but also anxiety and depressive like behaviours [87], which is of benefit in pain syndromes due to the close association of the stress and pain systems. These studies highlight the important dual-effect of the gut microbiota in disorders with both pain and affective components. Studies in germ-free mice show that inflammatory hypernociception induced by several different stimuli is reduced in germ-free mice [80]. These findings highlight the important role of the interaction between the commensal microbiota and the host in favouring adaptation to environmental stresses, including those that influence pain. Microbiota disruption with the Gram-negative bacterium *C. jejuni* leads to activation of brain regions that process gastrointestinal sensory information [88], while *Lactobacillus acidophilus* has been noted to modulate intestinal pain through opioid and cannabinoid receptors [89]. A study carried out in our laboratory has also indicated that probiotic bacteria is capable of reducing visceral pain in rodents [90]. An improvement in IBS symptoms in patients is also seen following the consumption of a variety of probiotic treatments [85]. Taken together, these studies indicate that visceral pain can be modulated through the gastrointestinal microbial system, but whether there is overlap or linkage with other systems that are more classically associated with pain and psychiatric disorders is yet to be elucidated.

Preclinical Studies Highlighting the Comorbidity of Visceral Pain and Psychiatric Disorders

Due to the intertwined nature of psychiatric disease and pain and given the comorbidity, several animal models encapsulate both of these elements. For some it was part of the design process and deliberate, while other animal models were initially designed to study one particular disease and subsequent study of the model uncovered that other systems were also affected.

The Maternal Separation Model

Maternal separation is a model of early-life stress [50, 91] that affects adult physiology and behaviour [92]. Early adverse events increase the vulnerability to development of disease such as depression, anxiety disorders and IBS [93, 94]. Indeed, patients with IBS often report a history of physical or sexual abuse along with a high rate of stressful events later in life [95]. Females are more susceptible to adverse early-life events and IBS [96, 97]. Despite this, most preclinical studies on the impact of early-life stress on visceral sensitivity employ male animals.

The maternally separated animal model has been widely studied in terms of brain-gut axis dysfunctions related to IBS. One of the most debilitating symptoms of this disorder is the increased perception of visceral stimuli. We and others have consistently shown that maternally separated animals develop visceral hypersensitivity in adulthood both under normal conditions and after stress [85, 96, 98–100]. IBS patients show elevated levels of psychological traits such as anxiety and depression [101–103]. Similarly, maternally separated animals display some phenotypic changes which are indicative of comorbid anxiety and depression in adulthood [50]. Colorectal distension leads to substantial increased blood flow in the periaqueductal grey, an area involved in descending pain modulation [104]. This model has also been associated with increased activation [105] and plasticity of neurons in the dorsal root ganglia in the lumbosacral part of the spinal cord [106].

In this model, offspring are separated from their mother during an important developmental postnatal period, the 'stress hyporesponsive period', which is considered important for maturation of the HPA axis in rats [107] and humans [108]. Brief separation of rat pups from their mother (i.e. 3–15 min) during the first 2–3 weeks of life typically decreases the magnitude of behavioural responses to stress and HPA axis activation in adulthood [91], and is often referred to as maternal deprivation. Conversely, long periods of separation (i.e. 3–6 h) from the mother during the postnatal period increase the magnitude of behavioural responses to stress and the HPA axis activation in adulthood [109, 110]. Nonetheless, this shorter period of separation leads to a sexually dimorphic thermal and mechanical allodynia [111]. Also spinal nerve ligation results in exacerbated mechanical and cold allodynia of the ipsilateral hind paw in female deprived pups only [111]. Early-life maternal separation has also been found to be associated with changes in the composition of the microbiota [99, 112]. These studies highlight the involvement of early-life stress in the dysfunction of the brain-gut-microbiota axis and the development of visceral pain and psychiatric disease.

The Wistar-Kyoto Model

The Wistar-Kyoto (WKY) is a strain of rat genetically predisposed to exaggerated stress responses characterised by alteration at the level of the HPA axis with exaggerated ACTH and corticosterone levels [113]. A number of behavioural paradigms indicate that these animals exhibit a depressive-like phenotype [114, 115]. Moreover, alteration in the central serotonergic system and alteration of sleep have also been noticed in these animals [116] as well as the endocannabinoid system [117, 118]. Because these characteristics parallel those seen in depressive patients, the WKY rat is considered a valid animal model of depression. Several studies have also demonstrated that WKY animals have increased sensitivity to colorectal distension [119, 120] coupled with alterations in colonic accommodation [121] and colonic morphology [122]. It has also been demonstrated that visceral hypersensitivity in the WKY rat is modulated by the central nucleus of the amygdala through the CRF_1 receptor [123]. A higher activation of the prefrontal cortex has also been found in WKY rats when compared to normosensitive animals in response to colorectal distension [124]. Evidence from our laboratory has also suggested that the serotonergic system through the 5-HT_{2B} receptor may be implicated in the increased visceral response observed in WKY rats [125]. Alteration in the expression of key elements of the immune systems such as Toll-like receptors have also been found in this animal model [49], correlating with human data [126]. The WKY rat allows the study of the involvement of a genetically higher stress response in the generation of psychiatric disease and visceral pain.

Water Avoidance Stress Model

In the water avoidance stress paradigm, the animal is placed for 1 h on a block in the centre of a tank filled with water to within 1 cm of the top of the block. This represents a potent psychological stressor, resulting in hypercorticosteronaemia [127]. Chronic water avoidance (10 days) induces persistent visceral hyperalgesia and these animals exhibit anxiety-like behaviours and enhanced faecal pellet excretion [127, 128]. In this animal model, visceral hypersensitivity has been associated with colonic immune activation [128]. Moreover, it has been suggested that chronic water avoidance stress-induced visceral hypersensitivity could be due to the downregulation of the endocannabinoid receptor CB1 (consistent with increased levels of anandamide in this animal model) at the level of the dorsal root ganglia (L6-S2) which, in turn, induces upregulation of transient receptor potential cation channel subfamily V member 1 expression and function [127]. Furthermore, others have shown that chronic water avoidance-induced effects are thought to be modulated at an epigenetic level, with HDAC inhibitors reversing water avoidance-induced hyperalgesia [129].

Restraint Stress

Restraint stress paradigms typically involve placing the animal in a well-ventilated tube or cage, which restricts but does not completely prevent movement [130, 131]. Exposure to partial restraint stress for 2 h was found to induce immediate hyperalgesia to colorectal distension in male [132] and female Wistar rats [133]. Moreover, 2 h of restraint per day over a 4-day period was also shown to induce visceral hyperalgesia in rats as measured by colorectal distension [134]. The time spent in an acute stress situation has proven crucial to the development of the IBS phenotype [128]. This model is regularly employed to study the behavioural and biological symptoms associated with human depressive disorders [54, 135, 136]. Recently, a study on prolonged restraint stress promoted depressive-like behaviour and altered cytokine profiles similar to those observed in major depressive disorder weeks after the stress was stopped [137]. Despite the advantages of this model, a lot of variations of the timings exist and do affect the response. For example, acute stressors of longer duration >2 h tend to be susceptible to the animals habituating to the stressful environment and this can confound the behavioural outputs.

Overlapping Aetiology of Visceral Pain and Psychiatric Disorders

The high comorbidity observed between visceral pain and psychiatric disorders suggests potential common neurobiological pathways involved in the aetiology of these disorders. It is now well accepted that alterations at the level of the serotonergic and noradrenergic pathways play a central role in the aetiology of depression [138]. Moreover, these systems are well recognised to modulate nociceptive transmission, and antidepressant drugs have demonstrated efficacy in the treatment of visceral pain [139]. The most common antidepressants used in clinical practice are tricyclic antidepressants, selective serotonin reuptake inhibitors and serotonin norepinephrine reuptake inhibitors [140]. These different classes of antidepressants have demonstrated efficacy in relieving abdominal pain both clinically and preclinically [139]. While duloxetine, a serotonin norepinephrine reuptake inhibitor, is a well-studied compound with regard to pain, it is of course not the only antidepressant effective in reducing pain [141]. Thus, antidepressant treatment may ameliorate visceral pain by improving the psychological and psychiatric symptoms as well as modulating visceral pain sensation and gut motility through the targeting of both the serotonergic and noradrenergic systems.

Ketamine has also been suggested to facilitate antidepressant actions through the blockade of the glutamate N-methyl-D-aspartate (NMDA) receptor [142]. A large body of evidence implicates the NMDA receptor in both central and peripheral sensitisation and the development of visceral hypersensitivity [143]. Moreover, pharmacological antagonism of spinal NMDA receptors has been shown to reduce visceral hypersensitivity [144]. To date, several groups have suggested the involvement of the gut microbiota in stress-related disorders [33, 64]. Preclinical work carried out in germ-free animals has revealed a relationship between gut microbiota and stress- and anxiety-related behaviours [85, 145]. Similarly, different probiotic treatments have also been shown to have beneficial effects on anxiety- and depression-like behaviours [66, 87, 146]. Interestingly, a number of studies have suggested the potential for the manipulation of the microbiota through probiotic administration as a new treatment for visceral pain [89, 147, 148]. Moreover, modifications of the microbiota, through the administration of antibiotics, have been associated with changes in visceral pain perception [77, 78, 86]. Interestingly, altered brain levels of serotonin and noradrenaline have been demonstrated following probiotic treatment, suggesting a mechanism by which microbiota may influence mood and pain perception [149].

Future Directions

The complex and multilayered communication between the gut microbial population and central nervous system has far reaching implications, not least in the area of visceral pain and psychiatric comorbidities. Being able to target one element of the brain-gut axis in order to alleviate symptoms at both ends is a huge therapeutic leap. Studies on the modulation of the gut microbiota via prebiotic/probiotic treatment have shone a very positive light on the potential for this therapy in comorbid visceral pain and psychiatric disorders [85, 90, 150, 151]. Also of potential therapeutic value is the possibility of faecal transplantation. Studies have shown that condition and behaviour are transferrable via transplantation of faeces, which may potentially offer the possibility to 'select' the desired phenotype by simply selecting a particular microbial community.

References

1 Cervero F, Laird JM: Visceral pain. Lancet 1999;353: 2145–2148.
2 Elsenbruch S, Rosenberger C, Enck P, Forsting M, Schedlowski M, Gizewski ER: Affective disturbances modulate the neural processing of visceral pain stimuli in irritable bowel syndrome: an fMRI study. Gut 2010;59:489–495.
3 Wessely S, Nimnuan C, Sharpe M: Functional somatic syndromes: one or many? Lancet 1999;354: 936–939.
4 Butler RK, Finn DP: Stress-induced analgesia. Prog Neurobiol 2009;88:184–202.
5 Jennings EM, Okine BN, Roche M, Finn DP: Stress-induced hyperalgesia. Prog Neurobiol 2014;121:1–18.
6 Olango WM, Finn DP: Neurobiology of stress-induced hyperalgesia. Curr Top Behav Neurosci 2014; 20:251–280.
7 Sikandar S, Dickenson AH: Visceral pain: the ins and outs, the ups and downs. Curr Opin Support Palliat Care 2012;6:17–26.
8 Cervero F: Visceral versus somatic pain: similarities and differences. Dig Dis 2009;27(suppl 1):3–10.
9 Ritchie J: Pain from distension of the pelvic colon by inflating a balloon in the irritable colon syndrome. Gut 1973;14:125–132.
10 Hungin AP, Whorwell PJ, Tack J, Mearin F: The prevalence, patterns and impact of irritable bowel syndrome: an international survey of 40,000 subjects. Aliment Pharmacol Ther 2003;17:643–650.
11 Handa M, Nukina H, Ando K, Kubo C: What does pain or discomfort in irritable bowel syndrome mean? Dig Dis Sci 2004;49:575–578.
12 Vercellini P, Somigliana E, Vigano P, Abbiati A, Barbara G, Fedele L: Chronic pelvic pain in women: etiology, pathogenesis and diagnostic approach. Gynecol Endocrinol 2009;25:149–158.
13 Prakash Gyawali C: Esophageal hypersensitivity. Gastroenterol Hepatol (NY) 2010;6:497–500.
14 Farmer AD, Coen SJ, Kano M, Paine PA, Shwahdi M, Jafari J, Kishor J, Worthen SF, Rossiter HE, Kumari V, Williams SC, Brammer M, Giampietro VP, Droney J, Riley J, Furlong PL, Knowles CH, Lightman SL, Aziz Q: Psychophysiological responses to pain identify reproducible human clusters. Pain 2013;154:2266–2276.
15 Giamberardino MA, Vecchiet L: Visceral pain, referred hyperalgesia and outcome: new concepts. Eur J Anaesthesiol Suppl 1995;10:61–66.
16 Fass R, Achem SR: Noncardiac chest pain: epidemiology, natural course and pathogenesis. J Neurogastroenterol Motil 2011;17:110–123.

17 Vedder H: Physiology of the hypothalamic-pituitary-adrenocortical axis; in del Rey GPC, Hugo OB (eds): NeuroImmune Biology. Amsterdam, Elsevier, 2007, vol 7, pp 17–31.
18 De Kloet ER, Vreugdenhil E, Oitzl MS, Joels M: Glucocorticoid feedback resistance. Trends Endocrinol Metab 1997;8:26–33.
19 McEwen BS: Mood disorders and allostatic load. Biol Psychiatry 2003;54:200–207.
20 Ladd CO, Huot RL, Thrivikraman KV, Nemeroff CB, Meaney MJ, Plotsky PM: Long-term behavioral and neuroendocrine adaptations to adverse early experience. Prog Brain Res 1999;122:81–103.
21 Zozulya AA, Gabaeva MV, Sokolov OY, Surkina ID, Kost NV: Personality, coping style, and constitutional neuroimmunology. J Immunotoxicol 2008;5:221–225.
22 Franklin TB, Saab BJ, Mansuy IM: Neural mechanisms of stress resilience and vulnerability. Neuron 2012;75:747–761.
23 Kessler RC, Gruber M, Hettema JM, Hwang I, Sampson N, Yonkers KA: Co-morbid major depression and generalized anxiety disorders in the National Comorbidity Survey follow-up. Psychol Med 2008;38:365–374.
24 Kessler RC, Bromet EJ: The epidemiology of depression across cultures. Annu Rev Public Health 2013;34:119–138.
25 Xiang X, An R: Depression and onset of cardiovascular disease in the US middle-aged and older adults. Aging Ment Health 2015, pp 1–9.
26 Dinan TG, Cryan JF: Melancholic microbes: a link between gut microbiota and depression? Neurogastroenterol Motil 2013;25:713–719.
27 Finn DP, Viveros MP, Marco EM: The endocannabinoid system and emotional processing: pathophysiology and therapeutic potential. J Psychopharmacol 2012;26:3–6.
28 Hornig M: The role of microbes and autoimmunity in the pathogenesis of neuropsychiatric illness. Curr Opin Rheumatol 2013;25:488–795.
29 Blanchard EB, Lackner JM, Jaccard J, Rowell D, Carosella AM, Powell C, Sanders K, Krasner S, Kuhn E: The role of stress in symptom exacerbation among IBS patients. J Psychosom Res 2008;64:119–128.
30 Théodorou V: Susceptibility to stress-induced visceral sensitivity: a bad legacy for next generations. Neurogastroenterol Motil 2013;25:927–930.
31 Sagami Y, Shimada Y, Tayama J, Nomura T, Satake M, Endo Y, Shoji T, Karahashi K, Hongo M, Fukudo S: Effect of a corticotropin releasing hormone receptor antagonist on colonic sensory and motor function in patients with irritable bowel syndrome. Gut 2004;53:958–964.
32 Simren M, Ringstrom G, Bjornsson ES, Abrahamsson H: Treatment with hypnotherapy reduces the sensory and motor component of the gastrocolonic response in irritable bowel syndrome. Psychosom Med 2004;66:233–238.
33 Mayer EA, Knight R, Mazmanian SK, Cryan JF: Gut microbes and the brain: paradigm shift in neuroscience. J Neurosci 2014;34:15490–15496.
34 Ji Y, Tang B, Cao DY, Wang G, Traub RJ: Sex differences in spinal processing of transient and inflammatory colorectal stimuli in the rat. Pain 2012;153:1965–1973.
35 Mogil JS: Sex differences in pain and pain inhibition: multiple explanations of a controversial phenomenon. Nat Rev Neurosci 2012;13:859–866.
36 Hashmi JA, Davis KD: Women experience greater heat pain adaptation and habituation than men. Pain 2009;145:350–357.
37 Widmalm SE, McKay DC, Radke JC, Zhang Y, Wang X, Wang M: Gender differences in low and high pain palpation thresholds in the TMJ and neck areas. Cranio 2013;31:92–99.
38 Mayer EA, Berman S, Chang L, Naliboff BD: Sex-based differences in gastrointestinal pain. Eur J Pain 2004;8:451–463.
39 Heitkemper MM, Chang L: Do fluctuations in ovarian hormones affect gastrointestinal symptoms in women with irritable bowel syndrome? Gend Med 2009;6(suppl 2):152–167.
40 Fillingim RB, King CD, Ribeiro-Dasilva MC, Rahim-Williams B, Riley JL 3rd: Sex, gender, and pain: a review of recent clinical and experimental findings. J Pain 2009;10:447–485.
41 Larauche M, Mulak A, Tache Y: Stress and visceral pain: from animal models to clinical therapies. Exp Neurol 2012;233:49–67.
42 Walker JS, Carmody JJ: Experimental pain in healthy human subjects: gender differences in nociception and in response to ibuprofen. Anesth Analg 1998;86:1257–1262.
43 Hong J-Y, Kilpatrick LA, Labus J, Gupta A, Jiang Z, Ashe-McNalley C, Stains J, Heendeniya N, Ebrat B, Smith S: Patients with chronic visceral pain show sex-related alterations in intrinsic oscillations of the resting brain. J Neurosci 2013;33:11994–12002.
44 Jiang Z, Dinov ID, Labus J, Shi Y, Zamanyan A, Gupta A, Ashe-McNalley C, Hong J-Y, Tillisch K, Toga AW: Sex-related differences of cortical thickness in patients with chronic abdominal pain. PLoS One 2013;8:e73932.
45 Mulak A, Tache Y, Larauche M: Sex hormones in the modulation of irritable bowel syndrome. World J Gastroenterol 2014;20:2433–2448.
46 Vilanova LS, Goncalves TM, Meirelles L, Garcia RC: Hormonal fluctuations intensify temporomandibular disorder pain without impairing masticatory function. Int J Prosthodont 2015;28:72–74.

47 Fatima N, Babu PR, Sisinty VS, Tarakji B: Pain perception and anxiety levels during menstrual cycle associated with periodontal therapy. Int J Dent 2014; 2014:472926.
48 Meseguer V, Alpizar YA, Luis E, Tajada S, Denlinger B, Fajardo O, Manenschijn JA, Fernandez-Pena C, Talavera A, Kichko T, Navia B, Sanchez A, Senaris R, Reeh P, Perez-Garcia MT, Lopez-Lopez JR, Voets T, Belmonte C, Talavera K, Viana F: TRPA1 channels mediate acute neurogenic inflammation and pain produced by bacterial endotoxins. Nat Commun 2014;5:3125.
49 McKernan DP, Nolan A, Brint EK, O'Mahony SM, Hyland NP, Cryan JF, Dinan TG: Toll-like receptor mRNA expression is selectively increased in the colonic mucosa of two animal models relevant to irritable bowel syndrome. PLoS One 2009;4:e8226.
50 O'Mahony SM, Marchesi JR, Scully P, Codling C, Ceolho AM, Quigley EM, Cryan JF, Dinan TG: Early life stress alters behavior, immunity, and microbiota in rats: implications for irritable bowel syndrome and psychiatric illnesses. Biol Psychiatry 2009;65: 263–267.
51 Hyland NP, Quigley EM, Brint E: Microbiota-host interactions in irritable bowel syndrome: epithelial barrier, immune regulation and brain-gut interactions. World J Gastroenterol 2014;20:8859–8866.
52 Nebot-Vivinus M, Harkat C, Bzioueche H, Cartier C, Plichon-Dainese R, Moussa L, Eutamene H, Pishvaie D, Holowacz S, Seyrig C, Piche T, Theodorou V: Multispecies probiotic protects gut barrier function in experimental models. World J Gastroenterol 2014;20:6832–6843.
53 Jones KA, Thomsen C: The role of the innate immune system in psychiatric disorders. Mol Cell Neurosci 2013;53:52–62.
54 Dantzer R, O'Connor JC, Freund GG, Johnson RW, Kelley KW: From inflammation to sickness and depression: when the immune system subjugates the brain. Nat Rev Neurosci 2008;9:46–56.
55 Levine J, Barak Y, Chengappa K, Rapoport A, Rebey M, Barak V: Cerebrospinal cytokine levels in patients with acute depression. Neuropsychobiology 1999; 40:171–176.
56 Maes M: Major depression and activation of the inflammatory response system; in Dantzer R, Wollmann E, Yirmiya R (eds): Cytokines, Stress, and Depression. Berlin, Springer, 1999, pp 25–46.
57 Hoge EA, Brandstetter K, Moshier S, Pollack MH, Wong KK, Simon NM: Broad spectrum of cytokine abnormalities in panic disorder and posttraumatic stress disorder. Depress Anxiety 2009;26:447–455.
58 Wang Z, Zhang Q, Yuan L, Wang S, Liu L, Yang X, Li G, Liu D: The effects of curcumin on depressive-like behavior in mice after lipopolysaccharide administration. Behav Brain Res 2014;274:282–290.
59 Grigoleit J-S, Kullmann JS, Wolf OT, Hammes F, Wegner A, Jablonowski S, Engler H, Gizewski E, Oberbeck R, Schedlowski M: Dose-dependent effects of endotoxin on neurobehavioral functions in humans. PLoS One 2011;6:e28330.
60 Tougas G: The autonomic nervous system in functional bowel disorders. Gut 2000;47(suppl 4):iv78–iv80; discussion iv87.
61 Mayer E, Tillisch K, Bradesi S: Review article: modulation of the brain – gut axis as a therapeutic approach in gastrointestinal disease. Aliment Pharmacol Ther 2006;24:919–933.
62 Bonaz BL, Bernstein CN: Brain-gut interactions in inflammatory bowel disease. Gastroenterology 2013; 144:36–49.
63 Moloney RD, Desbonnet L, Clarke G, Dinan TG, Cryan JF: The microbiome: stress, health and disease. Mamm Genome 2014;25:49–74.
64 Borre YE, Moloney RD, Clarke G, Dinan TG, Cryan JF: The impact of microbiota on brain and behavior: mechanisms & therapeutic potential. Adv Exp Med Biol 2014;817:373–403.
65 Gaykema RPA, Goehler LE, Lyte M: Brain response to cecal infection with *Campylobacter jejuni*: analysis with Fos immunohistochemistry. Brain Behav Immun 2004;18:238–245.
66 Bravo JA, Forsythe P, Chew MV, Escaravage E, Savignac HM, Dinan TG, Bienenstock J, Cryan JF: Ingestion of Lactobacillus strain regulates emotional behavior and central GABA receptor expression in a mouse via the vagus nerve. Proc Natl Acad Sci U S A 2011;108:16050–16055.
67 Sudo N, Chida Y, Aiba Y, Sonoda J, Oyama N, Yu XN, Kubo C, Koga Y: Postnatal microbial colonization programs the hypothalamic-pituitary-adrenal system for stress response in mice. J Physiol 2004; 558:263–275.
68 Neufeld KM, Kang N, Bienenstock J, Foster JA: Reduced anxiety-like behavior and central neurochemical change in germ-free mice. Neurogastroenterol Motil 2010;23:255–264.e119.
69 Heijtz RD, Wang S, Anuar F, Qian Y, Bjorkholm B, Samuelsson A, Hibberd ML, Forssberg H, Pettersson S: Normal gut microbiota modulates brain development and behavior. Proc Natl Acad Sci U S A 2011; 108:3047–3052.
70 Clarke G, Grenham S, Scully P, Fitzgerald P, Moloney RD, Shanahan F, Dinan TG, Cryan JF: The microbiome-gut-brain axis during early life regulates the hippocampal serotonergic system in a sex-dependent manner. Mol Psychiatry 2013;18:666–673.
71 Bercik P, Denou E, Collins J, Jackson W, Lu J, Jury J, Deng Y, Blennerhassett P, Macri J, McCoy KD, Verdu EF, Collins SM: The intestinal microbiota affect central levels of brain-derived neurotropic factor and behavior in mice. Gastroenterology 2011;141: 599–609, 609.e1–e3.

72 Desbonnet L, Garrett L, Clarke G, Bienenstock J, Dinan TG: The probiotic *Bifidobacteria infantis*: an assessment of potential antidepressant properties in the rat. J Psychiatr Res 2008;43:164–174.
73 Messaoudi M, Lalonde R, Violle N, Javelot H, Desor D, Nejdi A, Bisson JF, Rougeot C, Pichelin M, Cazaubiel M, Cazaubiel JM: Assessment of psychotropic-like properties of a probiotic formulation (Lactobacillus helveticus R0052 and *Bifidobacterium longum* R0175) in rats and human subjects. Br J Nutr 2011;105:755–764.
74 Miyaoka T, Wake R, Furuya M, Liaury K, Ieda M, Kawakami K, Tsuchie K, Taki M, Ishihara K, Araki T, Horiguchi J: Minocycline as adjunctive therapy for patients with unipolar psychotic depression: an open-label study. Prog Neuropsychopharmacol Biol Psychiatry 2012;37:222–226.
75 Soczynska JK, Mansur RB, Brietzke E, Swardfager W, Kennedy SH, Woldeyohannes HO, Powell AM, Manierka MS, McIntyre RS: Novel therapeutic targets in depression: minocycline as a candidate treatment. Behav Brain Res 2012;235:302–317.
76 Zemke D, Majid A: The potential of minocycline for neuroprotection in human neurologic disease. Clin Neuropharmacol 2004;27:293–298.
77 Aguilera M, Cerda-Cuellar M, Martinez V: Antibiotic-induced dysbiosis alters host-bacterial interactions and leads to colonic sensory and motor changes in mice. Gut Microbes 2015;6:10–23.
78 O'Mahony SM, Felice VD, Nally K, Savignac HM, Claesson MJ, Scully P, Woznicki J, Hyland NP, Shanahan F, Quigley EM, Marchesi JR, O'Toole PW, Dinan TG, Cryan JF: Disturbance of the gut microbiota in early-life selectively affects visceral pain in adulthood without impacting cognitive or anxiety-related behaviors in male rats. Neuroscience 2014;277:885–901.
79 Theodorou V, Ait Belgnaoui A, Agostini S, Eutamene H: Effect of commensals and probiotics on visceral sensitivity and pain in irritable bowel syndrome. Gut Microbes 2014;5:430–436.
80 Amaral FA, Sachs D, Costa VV, Fagundes CT, Cisalpino D, Cunha TM, Ferreira SH, Cunha FQ, Silva TA, Nicoli JR, Vieira LQ, Souza DG, Teixeira MM: Commensal microbiota is fundamental for the development of inflammatory pain. Proc Natl Acad Sci U S A 2008;105:2193–2197.
81 van Hemert S, Breedveld AC, Rovers JM, Vermeiden JP, Witteman BJ, Smits MG, de Roos NM: Migraine associated with gastrointestinal disorders: review of the literature and clinical implications. Front Neurol 2014;5:241.
82 McLean MH, Dieguez D Jr, Miller LM, Young HA: Does the microbiota play a role in the pathogenesis of autoimmune diseases? Gut 2015;64:332–341.
83 Smilowicz A: An osteopathic approach to gastrointestinal disease: somatic clues for diagnosis and clinical challenges associated with *Helicobacter pylori* antibiotic resistance. J Am Osteopath Assoc 2013; 113:404–416.
84 Collins SM: A role for the gut microbiota in IBS. Nat Rev Gastroenterol Hepatol 2014;11:497–505.
85 Clarke G, Cryan JF, Dinan TG, Quigley EM: Review article: probiotics for the treatment of irritable bowel syndrome – focus on lactic acid bacteria. Aliment Pharmacol Ther 2012;35:403–413.
86 Verdu EF, Bercik P, Verma-Gandhu M, Huang XX, Blennerhassett P, Jackson W, Mao Y, Wang L, Rochat F, Collins SM: Specific probiotic therapy attenuates antibiotic induced visceral hypersensitivity in mice. Gut 2006;55:182–190.
87 Desbonnet L, Garrett L, Clarke G, Kiely B, Cryan JF, Dinan TG: Effects of the probiotic *Bifidobacterium infantis* in the maternal separation model of depression. Neuroscience 2010;170:1179–1188.
88 Gaykema RP, Goehler LE, Lyte M: Brain response to cecal infection with *Campylobacter jejuni:* analysis with Fos immunohistochemistry. Brain Behav Immun 2004;18:238–245.
89 Rousseaux C, Thuru X, Gelot A, Barnich N, Neut C, Dubuquoy L, Dubuquoy C, Merour E, Geboes K, Chamaillard M, Ouwehand A, Leyer G, Carcano D, Colombel JF, Ardid D, Desreumaux P: *Lactobacillus acidophilus* modulates intestinal pain and induces opioid and cannabinoid receptors. Nat Med 2007;13: 35–37.
90 McKernan DP, Fitzgerald P, Dinan TG, Cryan JF: The probiotic *Bifidobacterium infantis* 35624 displays visceral antinociceptive effects in the rat. Neurogastroenterol Motil 2010;22:1029–1035.e268.
91 Levine S: Enduring effects of early experience on adult behavior; in Donald WP, et al (eds): Hormones, Brain and Behavior. San Diego, Academic Press, 2002, pp 535–542.
92 Plotsky PM, Thrivikraman KV, Nemeroff CB, Caldji C, Sharma S, Meaney MJ: Long-term consequences of neonatal rearing on central corticotropin-releasing factor systems in adult male rat offspring. Neuropsychopharmacology 2005;30:2192–2204.
93 Heim C, Newport DJ, Wagner D, Wilcox MM, Miller AH, Nemeroff CB: The role of early adverse experience and adulthood stress in the prediction of neuroendocrine stress reactivity in women: a multiple regression analysis. Depress Anxiety 2002;15:117–125.
94 Folks DG: The interface of psychiatry and irritable bowel syndrome. Curr Psychiatry Rep 2004;6:210–215.
95 Gaynes BN, Drossman DA: The role of psychosocial factors in irritable bowel syndrome. Baillieres Best Pract Res Clin Gastroenterol 1999;13:437–452.

96 Gue M, Del Rio-Lacheze C, Eutamene H, Theodorou V, Fioramonti J, Bueno L: Stress-induced visceral hypersensitivity to rectal distension in rats: role of CRF and mast cells. Neurogastroenterol Motil 1997;9:271–279.
97 Bradford K, Shih W, Videlock EJ, Presson AP, Naliboff BD, Mayer EA, Chang L: Association between early adverse life events and irritable bowel syndrome. Clin Gastroenterol Hepatol 2012;10: 385–390.e383.
98 Coutinho SV, Plotsky PM, Sablad M, Miller JC, Zhou H, Bayati AI, McRoberts JA, Mayer EA: Neonatal maternal separation alters stress-induced responses to viscerosomatic nociceptive stimuli in rat. Am J Physiol Gastrointest Liver Physiol 2002; 282:G307–G316.
99 O'Mahony SM, Marchesi JR, Scully P, Codling C, Ceolho AM, Quigley EM, Cryan JF, Dinan TG: Early life stress alters behavior, immunity, and microbiota in rats: implications for irritable bowel syndrome and psychiatric illnesses. Biol Psychiatry 2009;65:263–267.
100 O'Mahony SM, Hyland NP, Dinan TG, Cryan JF: Maternal separation as a model of brain-gut axis dysfunction. Psychopharmacology (Berl) 2011; 214:71–88.
101 Whitehead WE, Engel BT, Schuster MM: Irritable bowel syndrome: physiological and psychological differences between diarrhea-predominant and constipation-predominant patients. Dig Dis Sci 1980;25:404–413.
102 Garakani A, Win T, Virk S, Gupta S, Kaplan D, Masand PS: Comorbidity of irritable bowel syndrome in psychiatric patients: a review. Am J Ther 2003; 10:61–67.
103 Clarke G, Quigley EM, Cryan JF, Dinan TG: Irritable bowel syndrome: towards biomarker identification. Trends Mol Med 2009;15:478–489.
104 Wouters MM, Van Wanrooy S, Casteels C, Nemethova A, de Vries A, Van Oudenhove L, Van den Wijngaard RM, Van Laere K, Boeckxstaens G: Altered brain activation to colorectal distention in visceral hypersensitive maternal-separated rats. Neurogastroenterol Motil 2012;24:678–685.e297.
105 Zhang XJ, Li Z, Chung EK, Zhang HQ, Xu HX, Sung JJ, Bian ZX: Activation of extracellular signal-regulated protein kinase is associated with colorectal distension-induced spinal and supraspinal neuronal response and neonatal maternal separation-induced visceral hyperalgesia in rats. J Mol Neurosci 2009;37: 274–287.
106 Chung EK, Zhang XJ, Xu HX, Sung JJ, Bian ZX: Visceral hyperalgesia induced by neonatal maternal separation is associated with nerve growth factor-mediated central neuronal plasticity in rat spinal cord. Neuroscience 2007;149:685–695.
107 Walker CD, Perrin M, Vale W, Rivier C: Ontogeny of the stress response in the rat: role of the pituitary and the hypothalamus. Endocrinology 1986;118: 1445–1451.
108 Gunnar MR, Donzella B: Social regulation of the cortisol levels in early human development. Psychoneuroendocrinology 2002;27:199–220.
109 Plotsky PM, Meaney MJ: Early, postnatal experience alters hypothalamic corticotropin-releasing factor (CRF) mRNA, median eminence CRF content and stress-induced release in adult rats. Mol Brain Res 1993;18:195–200.
110 Ladd CO, Owens MJ, Nemeroff CB: Persistent changes in corticotropin-releasing factor neuronal systems induced by maternal deprivation. Endocrinology 1996;137:1212–1218.
111 Burke NN, Llorente R, Marco EM, Tong K, Finn DP, Viveros MP, Roche M: Maternal deprivation is associated with sex-dependent alterations in nociceptive behavior and neuroinflammatory mediators in the rat following peripheral nerve injury. J Pain 2013;14:1173–1184.
112 Bailey MT, Dowd SE, Parry NM, Galley JD, Schauer DB, Lyte M: Stressor exposure disrupts commensal microbial populations in the intestines and leads to increased colonization by *Citrobacter rodentium*. Infect Immun 2010;78:1509–1519.
113 Solberg LC, Olson SL, Turek FW, Redei E: Altered hormone levels and circadian rhythm of activity in the WKY rat, a putative animal model of depression. Am J Physiol Regul Integr Comp Physiol 2001;281:R786–R794.
114 Paré WP: Open field, learned helplessness, conditioned defensive burying, and forced-swim tests in WKY rats. Physiol Behav 1994;55:433–439.
115 Lahmame A, del Arco C, Pazos A, Yritia M, Armario A: Are Wistar-Kyoto rats a genetic animal model of depression resistant to antidepressants? Eur J Pharmacol 1997;337:115–123.
116 Dugovic C, Solberg LC, Redei E, Van Reeth O, Turek FW: Sleep in the Wistar-Kyoto rat, a putative genetic animal model for depression. Neuroreport 2000;11:627–631.
117 Rea K, Olango WM, Okine BN, Madasu MK, McGuire IC, Coyle K, Harhen B, Roche M, Finn DP: Impaired endocannabinoid signalling in the rostral ventromedial medulla underpins genotype-dependent hyper-responsivity to noxious stimuli. Pain 2014;155:69–79.
118 Vinod KY, Xie S, Psychoyos D, Hungund BL, Cooper TB, Tejani-Butt SM: Dysfunction in fatty acid amide hydrolase is associated with depressive-like behavior in Wistar Kyoto rats. PLoS One 2012;7: e36743.

119 Gunter WD, Shepard JD, Foreman RD, Myers DA, Greenwood-Van Meerveld B: Evidence for visceral hypersensitivity in high-anxiety rats. Physiol Behav 2000;69:379–382.
120 Greenwood-Van Meerveld B, Johnson AC, Cochrane S, Schulkin J, Myers DA: Corticotropin-releasing factor 1 receptor-mediated mechanisms inhibit colonic hypersensitivity in rats. Neurogastroenterol Motil 2005;17:415–422.
121 Martinez V, Ryttinger M, Kjerling M, Astin-Nielsen M: Characterisation of colonic accommodation in Wistar Kyoto rats with impaired gastric accommodation. Naunyn Schmiedebergs Arch Pharmacol 2007;376:205–216.
122 O'Malley D, Julio-Pieper M, Gibney SM, Dinan TG, Cryan JF: Distinct alterations in colonic morphology and physiology in two rat models of enhanced stress-induced anxiety and depression-like behaviour. Stress 2010;13:114–122.
123 Johnson AC, Tran L, Schulkin J, Meerveld BG-V: Importance of stress receptor-mediated mechanisms in the amygdala on visceral pain perception in an intrinsically anxious rat. Neurogastroenterol Motil 2012;24:479–486.
124 Gibney SM, Gosselin RD, Dinan TG, Cryan JF: Colorectal distension-induced prefrontal cortex activation in the Wistar-Kyoto rat: implications for irritable bowel syndrome. Neuroscience 2010;165: 675–683.
125 O'Mahony S, Bulmer D, Coelho A, Fitzgerald P, Bongiovanni C, Lee K, Winchester W, Dinan T, Cryan J: 5-HT_{2B} receptors modulate visceral hypersensitivity in a stress-sensitive animal model of brain-gut axis dysfunction. Neurogastroenterol Motility 2010;22:573–578.e124.
126 McKernan DP, Gaszner G, Quigley EM, Cryan JF, Dinan TG: Altered peripheral Toll-like receptor responses in the irritable bowel syndrome. Aliment Pharmacol Ther 2011;33:1045–1052.
127 Hong S, Fan J, Kemmerer ES, Evans S, Li Y, Wiley JW: Reciprocal changes in vanilloid (TRPV1) and endocannabinoid (CB1) receptors contribute to visceral hyperalgesia in the water avoidance stressed rat. Gut 2009;58:202–210.
128 Bradesi S, Schwetz I, Ennes HS, Lamy CM, Ohning G, Fanselow M, Pothoulakis C, McRoberts JA, Mayer EA: Repeated exposure to water avoidance stress in rats: a new model for sustained visceral hyperalgesia. Am J Physiol Gastrointest Liver Physiol 2005;289:G42–G53.
129 Tran L, Chaloner A, Sawalha AH, Greenwood Van-Meerveld B: Importance of epigenetic mechanisms in visceral pain induced by chronic water avoidance stress. Psychoneuroendocrinology 2013; 38:898–906.
130 da Silva Torres IL, Cucco SN, Bassani M, Duarte MS, Silveira PP, Vasconcellos AP, Tabajara AS, Dantas G, Fontella FU, Dalmaz C, Ferreira MB: Long-lasting delayed hyperalgesia after chronic restraint stress in rats-effect of morphine administration. Neurosci Res 2003;45:277–283.
131 Bardin L, Malfetes N, Newman-Tancredi A, Depoortere R: Chronic restraint stress induces mechanical and cold allodynia, and enhances inflammatory pain in rat: relevance to human stress-associated painful pathologies. Behav Brain Res 2009;205:360–366.
132 Gue M, Del Rio-Lacheze C, Eutamene H, Theodorou V, Fioramonti J, Bueno L: Stress-induced visceral hypersensitivity to rectal distension in rats: role of CRF and mast cells. Neurogastroenterol Motil 1997;9:271–279.
133 Rosztoczy A, Fioramonti J, Jarmay K, Barreau F, Wittmann T, Bueno L: Influence of sex and experimental protocol on the effect of maternal deprivation on rectal sensitivity to distension in the adult rat. Neurogastroenterol Motil 2003;15:679–686.
134 Eutamene H, Bradesi S, Larauche M, Theodorou V, Beaufrand C, Ohning G, Fioramonti J, Cohen M, Bryant AP, Kurtz C, Currie MG, Mayer EA, Bueno L: Guanylate cyclase C-mediated antinociceptive effects of linaclotide in rodent models of visceral pain. Neurogastroenterol Motil 2010;22:312.e84.
135 Boyle MP, Brewer JA, Funatsu M, Wozniak DF, Tsien JZ, Izumi Y, Muglia LJ: Acquired deficit of forebrain glucocorticoid receptor produces depression-like changes in adrenal axis regulation and behavior. Proc Natl Acad Sci U S A 2005;102: 473–478.
136 Maccari S, Morley-Fletcher S: Effects of prenatal restraint stress on the hypothalamus-pituitary-adrenal axis and related behavioural and neurobiological alterations. Psychoneuroendocrinology 2007;32(suppl 1):S10–S15.
137 Voorhees JL, Tarr AJ, Wohleb ES, Godbout JP, Mo X, Sheridan JF, Eubank TD, Marsh CB: Prolonged restraint stress increases IL-6, reduces IL-10, and causes persistent depressive-like behavior that is reversed by recombinant IL-10. PLoS One 2013; 8:e58488.
138 Stahl SM: Essential Psychopharmacology of Depression and Bipolar Disorder. Cambridge, Cambridge University Press, 2000.
139 Ford AC, Talley NJ, Schoenfeld PS, Quigley EM, Moayyedi P: Efficacy of antidepressants and psychological therapies in irritable bowel syndrome: systematic review and meta-analysis. Gut 2009;58: 367–378.
140 Lopez-Munoz F, Alamo C: Monoaminergic neurotransmission: the history of the discovery of antidepressants from 1950s until today. Curr Pharm Des 2009;15:1563–1586.

141 Kim SC, Landon JE, Lee YC: Patterns of health care utilization related to initiation of amitriptyline, duloxetine, gabapentin or pregabalin in fibromyalgia. Arthritis Res Ther 2015;17:18.
142 Monteggia LM, Zarate C Jr: Antidepressant actions of ketamine: from molecular mechanisms to clinical practice. Curr Opin Neurobiol 2015;30c:139–143.
143 Petrenko AB, Yamakura T, Baba H, Shimoji K: The role of N-methyl-D-aspartate (NMDA) receptors in pain: a review. Anesth Analg 2003;97:1108–1116.
144 Lin C, Al-Chaer ED: Differential effects of glutamate receptor antagonists on dorsal horn neurons responding to colorectal distension in a neonatal colon irritation rat model. World J Gastroenterol 2005;11:6495.
145 Neufeld KM, Kang N, Bienenstock J, Foster JA: Reduced anxiety-like behavior and central neurochemical change in germ-free mice. Neurogastroenterol Motil 2011;23:255–264.e119.
146 Bercik P, Park A, Sinclair D, Khoshdel A, Lu J, Huang X, Deng Y, Blennerhassett P, Fahnestock M, Moine D: The anxiolytic effect of *Bifidobacterium longum* NCC3001 involves vagal pathways for gut-brain communication. Neurogastroenterol Motil 2011;23:1132–1139.
147 Agrawal A, Houghton LA, Morris J, Reilly B, Guyonnet D, Goupil Feuillerat N, Schlumberger A, Jakob S, Whorwell PJ: Clinical trial: the effects of a fermented milk product containing *Bifidobacterium lactis* DN-173 010 on abdominal distension and gastrointestinal transit in irritable bowel syndrome with constipation. Aliment Pharmacol Ther 2009;29:104–114.
148 Agostini S, Goubern M, Tondereau V, Salvador-Cartier C, Bezirard V, Leveque M, Keranen H, Theodorou V, Bourdu-Naturel S, Goupil-Feuillerat N, Legrain-Raspaud S, Eutamene H: A marketed fermented dairy product containing *Bifidobacterium lactis* CNCM I-2494 suppresses gut hypersensitivity and colonic barrier disruption induced by acute stress in rats. Neurogastroenterol Motil 2012;24:376.e172.
149 Kannampalli P, Pochiraju S, Chichlowski M, Berg BM, Rudolph C, Bruckert M, Miranda A, Sengupta JN: Probiotic *Lactobacillus rhamnosus* GG (LGG) and prebiotic prevent neonatal inflammation-induced visceral hypersensitivity in adult rats. Neurogastroenterol Motil 2014;26:1694–1704.
150 Johnson AC, Greenwood-Van Meerveld B, McRorie J: Effects of *Bifidobacterium infantis* 35624 on post-inflammatory visceral hypersensitivity in the rat. Dig Dis Sci 2011;56:3179–3186.
151 Dai C, Guandalini S, Zhao DH, Jiang M: Antinociceptive effect of VSL#3 on visceral hypersensitivity in a rat model of irritable bowel syndrome: a possible action through nitric oxide pathway and enhance barrier function. Mol Cell Biochem 2012;362:43–53.
152 Canavan C, West J, Card T: The epidemiology of irritable bowel syndrome. Clin Epidemiol 2014;6:71–80.
153 Lacy B, Talley N, Locke Gr, Bouras E, DiBaise J, El-Serag H, Abraham B, Howden C, Moayyedi P, Prather C: Review article: current treatment options and management of functional dyspepsia. Aliment Pharmacol Ther 2012;36:3–15.
154 Farmer AD, Aziz Q: Mechanisms and management of functional abdominal pain. J R Soc Med 2014;107:347–354.
155 Bulun SE: Endometriosis. N Engl J Med 2009;360:268–279.
156 McLennan MT: Interstitial cystitis: epidemiology, pathophysiology, and clinical presentation. Obstet Gynecol Clin North Am 2014;41:385–395.
157 Krieger JN, Lee SWH, Jeon J, Cheah PY, Liong ML, Riley DE: Epidemiology of prostatitis. Int J Antimicrob Agents 2008;31:85–90.
158 Klinkman MS, Stevens D, Gorenflo DW: Episodes of care for chest pain: a preliminary report from MIRNET. Michigan Research Network. J Fam Pract 1994;38:345–352.
159 Park JJ, Park SJ, Choi DJ: Microvascular angina: angina that predominantly affects women. Korean J Intern Med 2015;30:140–147.
160 Walker EA, Katon WJ, Jemelka RP, Roy-Byrne PP: Comorbidity of gastrointestinal complaints, depression, and anxiety in the epidemiologic catchment area (ECA) study. Am J Med 1992;92:S26–S30.

Dr. Siobhain M. O'Mahony
Department of Anatomy and Neuroscience, University College
Western Gateway Building, Office 4.09
Cork (Ireland)
E-Mail somahony@ucc.ie

Finn DP, Leonard BE (eds): Pain in Psychiatric Disorders. Mod Trends Pharmacopsychiatry.
Basel, Karger, 2015, vol 30, pp 120–137 (DOI: 10.1159/000435937)

Genetic and Epigenetic Mechanisms Linking Pain and Psychiatric Disorders

Artur H. Swiergiel[a, c] • Grzegorz R. Juszczak[b] • Adrian M. Stankiewicz[b]

[a]Department of Animal and Human Physiology, Gdansk University, Gdansk, and [b]Department of Animal Behavior, Institute of Genetics and Animal Breeding, Polish Academy of Sciences, Jastrzebiec, Poland; [c]Department of Pharmacology, Toxicology and Neuroscience, Louisiana State University Health Sciences Center, Shreveport, La., USA

Abstract

The neurophysiological link between neuropathic pain and depression remains unknown despite evident high comorbidity of these two disorders. However, there is convincing evidence that genotype plays a role in both pain and depression. Using various types of genetic analysis – population genetics, cytogenetics and molecular technologies – specific genes have been implicated in mediating almost all aspects of nociception and mood disorders. The current review attempts to identify specific genes and epigenetic mechanisms common to both disorders. It is concluded that external and internal factors (inflammation, stress, gender, etc.) that contribute to the pathologies may do so through epigenetic mechanisms that may affect expression of these particular genes. The possible involvement of epigenetic regulation in pain and psychiatric disorders suggests that treatments targeting epigenetic mechanisms that mediate adverse life events should be considered.

The relationships between (different) pain(s) and psychiatric disorders have been thoroughly discussed in the previous chapters. Depression and pain comorbidity has also been well documented [1–3]. The question now arises whether and which genetic and epigenetic mechanisms are involved in mediating pain and/or depression. Furthermore, there is a need to discuss the significance of genotype and the epigenome with regard to depression and chronic pain comorbidity and whether any pain × psychiatric disorder relationship or comorbidity is affected by genetic or epigenetic mechanisms. Animal, human and translational research presented and discussed in excellent comprehensive reviews has suggested the significance of genotype in mediating pain [4–6], depression [7–10] and, through the shared genetic determinants, their comorbidity [11, 12]. Several reviews are devoted to the role of epigenetics in pain [13–15] and psychiatric disorders [16–18]. Yet, there have been but few attempts to identify and implicate common epigenetic mechanisms in both pain and psychiatric disorders [15, 19]. Here we strive to describe genetic and epigenetic mechanisms that could link pain and depression.

Genetics of Pain

There is a large variability in pain threshold or sensitivity and analgesia in humans [20, 21] and animals [22, 23]. Although selective breeding programs have revealed that pain sensitivity is a highly heritable trait [5, 22, 24], it has been extremely difficult to identify genetic mechanisms, not to mention specific genes, underlying these differences [4, 25]. Even in the case of opioid analgesia mediated by a simple biological mechanism, there is rather weak evidence for an association between effective opioid dosage requirements and mu-opioid receptor gene polymorphism [26]. Much more progress has been obtained in the case of human hereditary pain syndromes associated either with insensitivity to pain or with experience of exaggerated pain [27]. These studies have revealed 13 genes that exert significant effects on pain sensitivity [27]. Interestingly, genes coding for opioid receptors – the most obvious candidates – have not been implicated in extreme pain phenotypes in humans. Instead, genes underlying human hereditary pain syndromes code for ion channels, enzymes and trophic factors, or have other unknown biological functions [27].

In addition to the very few genes exerting strong effects on pain sensitivity, there are a large number of genes displaying relatively weak effects [27]. Combination of these common alleles is largely responsible for individual differences in pain sensitivity [27]. A recent review identified 410 genes contributing to pain sensitivity in humans and animals [27]. These genes belong to 12 functional categories, which include ion transport and homeostasis, synaptic transmission, intracellular signal transduction, injury response, and anatomical development [27] [see also Pain Genes Database (PGdb); http://www.jbldesign.com/jmogil/enter.html [28]]. Among the listed genes, those coding for various cytokines (interleukins, interferon-gamma, tumor necrosis factor) and their receptors are especially interesting because cytokines have been implicated both in pain mechanisms [27] and in the etiology of mood disorders [29–32] (see the chapters by Lees et al. [this vol., pp. 51–66] and Madasu et al. [this vol., pp. 80–93].

Pain and depression may also be linked by genes involved in stress responses that are, at least in part, genetically determined [33–35]. Stress is a potent modulator of pain sensitivity [36–39] and also contributes to the mechanisms of mood disorders [40–42] (see the chapter by Leonard [this vol., pp. 22–35]).

Another important aspect may be sex-specific genetic mechanisms underlying differences between males and females in pain sensitivity [43, 44] because sex affects not only pain sensation [45, 46], but also predisposes to mood disorders [47, 48] and ultimately may play a role in pain × depression comorbidity [49, 50]. Even more importantly, there is a genetic link between sex and stress-induced differences in pain sensitivity [51–53].

Therefore, there are several factors, such as inflammation, stress and sex, that have been implicated in pain regulation and mood disorders, and may share common mechanisms and genetic underpinnings. Having delineated the potential role of genes in pain, it should be considered that their functional significance may depend on their expression. To a large extent, this depends on epigenetic effects (epigenetic mechanisms). It is thus highly plausible that some of these genes can be regulated by epigenetic processes [13, 14, 35, 54].

Genetics of Depression

Family, twin and adoption research and mendelian genetics methods have established that genes play a substantial role in psychiatric disorders, including depression [7–10, 55]. Additionally, genetic analysis involving molecular biology methods have indicated the contribution of multiple genes to a complex disorder such as depression. Several genome-wide association studies linking single-nucleotide polymorphism(s) (SNP) with a presumably well-defined and diagnosed (depressive) trait (symptom) and involving tens of thousands of cases have tried to determine the genetic etiology of depression [56]. It has to be recognized that despite all efforts to date, no single gene or genetic variation that increases the risk of depression substantially has been identified [57]. The most likely reason for these results is the particularly high heterogeneity of depressive disorders. Genetic variants exert only small effects on overall disease risk, and multiple genetic factors interacting with environmental factors are likely necessary for the development of depression. Furthermore, there is some lack of consistency in association studies, in both candidate gene and genome-wide association studies. Numerous factors may account for variable results, including the use of different diagnostic approaches, small samples in early studies, population stratification, copy number variation, rare variation, and phenotypic and allelic heterogeneity. The conflicting results are also probably, in part, a consequence of environmental factors and epigenetic effects not being considered or controlled for [7]. Nevertheless, replicated findings in the literature have suggested a prominent role in depression for several genes (though a high rate of failure of replication has also been reported).

In humans, the promoter region of the gene coding for the serotonin transporter (*SLC6A4*) harbors a 43-base pair insertion/deletion referred to as the 5-HTT-linked polymorphic region (5-HTTLPR). This polymorphism has been associated with a number of affective disorders including depression [58, 59]. In the case of this gene, of particular interest is the possibility of a gene × environment interaction between the polymorphic region and stressful life events [58, 60]. Furthermore, there is a possibility of 5-HTTLPR modulating the association between stress and depression [61]. Finally, there is some evidence for epigenetic regulation of 5-HTT expression and behavior in infant rhesus macaques [62].

Among other genes implicated in depression, the most prominent are those coding for serotonin receptor 1A (HTR1A), serotonin receptor 2A (HTR2A), brain-derived neurotrophic factor (BDNF), corticotropin-releasing hormone receptor 1 (CRHR1) and FK506-binding protein 5 (FKBP5) [7, 9]. Genes coding for opioid receptors are of some interest because of the suggested role of the opioid system in depressive phenotypes [63–65]. Inconsistent results have been reported for genes coding catechol-O-methyl transferase (COMT), which degrades catecholamines and other genes listed in dedicated databases of depression-related genes [66].

Common Genes and Genetic Mechanisms for Pain and Depression

There is some indirect evidence (provided by clinical pharmacotherapy and preclinical animal studies) that there may be common genetic determinants of pain and depression [2]. Clinical data clearly suggest close intertwining between mood regulation and pain perception at the levels of the central and peripheral nervous systems and the neurotransmitter systems [1, 3]. First of all, etiological hypotheses explaining the coexistence of pain and depression have been based on the demonstrated dysfunctions of the serotonergic and noradrenergic systems involved not only in mood control, but also in control of ascending pain stimuli [49]. Antidepressant drugs result in a normalization of the activity of those neurotransmitter systems. Dual-action drugs, affecting both serotonergic and noradrenergic systems, such as tricyclic antidepressants and venlafaxine, milnacipran, duloxetine and mirtazapine have been shown to alleviate both pain and depressive symptoms [67]. Extended use of these drugs in treating both disorders suggests some common underlying mechanisms. However, there are very few reports of specific genes clearly implicated in symptoms of both pain and depression although, for example, an association of serotonin-1A and -2A receptor promoter polymorphisms with depressive symptoms and pain (after lumbar disc surgery) has been described [68].

As far as preclinical animal studies are concerned, it was observed that lines of mice differing in stress-induced analgesia and then in responses to morphine and naloxone differed significantly in depression-like behaviors [69–71]. The roles of opioid receptors in pain have been extensively studied, but their function in mood disorders has received less attention. Nevertheless, evidence from animal research has revealed that opioid receptors exert highly distinct controls over mood-related processes via functional interactions between opioid receptors and monoaminergic systems relevant to mood control [64, 72]. Indeed, it was reported that behavioral responses to antidepressants (NK3 tachykinin receptor agonist aminosenktide [73]; desipramine, venlafaxine [71, 74]) were different between low stress-induced analgesia and high stress-induced analgesia lines. Importantly, gene expression was different between the lines, both in untreated and desipramine-treated mice [75, 76]. More preclinical evidence for the causative relationship between pain and depression has been recently reviewed [77]. Still, the shared genetic background of pain and depression is not so obvious and easy to decide on. Genetic variation in genes across the serotonergic system [mainly polymorphism in serotonin receptor 2A (*HTR2A*) and transporter] and adrenergic system (already implicated in depression) has also been associated with pain processes [78–80].

Other reports, however, should also be considered [33, 34]. Though the authors reported an association of SNPs in *HTR2A*, *SERPINA6* and *TPH2* with somatic symptoms, implicating them as potentially important in the shared genetic component of functional somatic syndromes, they also suggested that the results should be critically assessed (see [81]). Burke et al. [82] reported common changes in the expression of neuroinflammatory genes in the amygdala in a rat model of depression and neuropathic pain. However, the shared gene responses were restricted specifically to thermal but not mechanical stimuli. Furthermore, the responses were specific to particular models of pain and depression (spinal nerve ligation and olfactory bulbectomy). Another example are mice in which constitutive ablation of the *Trek1* gene [responsible for K(2P) channels mediating background K(+) conductance] correlates with increased sensitivity to thermal and mechanical pain, as well as with resistance, rather than susceptibility, to depression [83]. Nevertheless, shared commonalities between pain and depression at least suggest a possibility of common underlying genetic mechanisms.

Identifying Common Genes and Genetic Mechanisms

Due to the enormous and constantly growing body of literature on depression and pain, the classical method of typing keywords into search engines connected to literature databases, such as PubMed, is no longer practicable as it results in long lists of article titles and abstracts. Fortunately, user-friendly tools are now available that allow researchers who are not so computer savvy to speed up this process. With advances in algorithms, databases and software, more reliable and practical tools for mining connections between biological terms (defined as complete chains of events or molecular functions leading to a phenotype) are being developed. Currently, one can take advantage of several web data mining tools (table 1). These tools utilize different methods for gene prioritization. For example Gene2MeSH [84] (http://gene2mesh.ncibi.org), SNPs3d [85] (http://www.snps3d.org) and Coremine (http://www.coremine.com/medical/#search) search for co-occurrence of a given term (e.g. a disease) and genes in the titles and abstracts included in the MEDLINE or Pubmed databases. On the other hand, tools such as BioGraph [86] (http://www.biograph.be) or neXtProt [87] (http://www.nextprot.org) integrate information already present in an outside database, such as OMIM, miR2Disease, CTD, IntAct, KEGG, GO, and others for Biograph, or UniProt Swiss-Prot for neXtProt. Due to usage of different algorithms, databases and different interpretations of query formulation by a user (table 1), results from

Table 1. Interpretation of queries: depression, pain by different data mining tools

Data mining tool	Query type	Depression	Pain
BioGraph	Unified Medical Language System concepts	Depressive disorder; mental depression	Pain
Coremine	Complex vocabulary, i.e. MeSH and GO	Depression (MeSH)	Pain (symptom)
Gene2MeSH	MeSH concepts	Depression; depressive disorder; depressive disorder, major	Pain; pain, intractable; pain, postoperative; pain threshold; pain insensitivity, congenital
neXtProt	Complex vocabulary, i.e. MeSH and GO	Depression	Pain
SNPs3d	SNPs3d-specified MEDLINE query	Depression = depression: major depression; depressive symptoms; depressive; depressive disorder, major; antidepressant; antidepressant treatment; Hamilton; depressive disorder; life events; antidepressive agents; depressed patients; MDD; stressful life; antidepressant response; depression rating; major depressive; depression scale; unipolar depression; anxiety; Hamilton depression; antidepressants; unipolar; Hamilton rating; geriatric depression; disorder MDD; selective serotonin; serotonin reuptake; serotonin uptake inhibitors; Hamilton rating scale; mood; major depressive disorder; reuptake inhibitors; anxiety disorders; rating scale; depressive disorders; depression inventory; depression anxiety; major depressive disorder MDD; depression background	Pain = pain; CIPA; abdominal pain; chest pain; pain sensitivity; neuropathic pain; anhidrosis; pelvic pain; pain perception; pain threshold; analgesic; neuropathic; pain measurement; nociceptors; analgesia; hyperalgesia; chronic pain; nociceptive; congenital insensitivity; painful; back pain; pain syndrome; analgesics; inflammatory pain; nociception; severe pain; TRPV1; analgesics, opioid; muscle pain; capsaicin; pain relief; capsaicin receptor; scn9a protein, human; neuralgia; noxious; chest; acute pain; noxious stimuli; pain, postoperative; pain insensitivity, congenital

BioGraph: integrates heterogeneous knowledge bases in case of disease-gene association – OMIM, miR2Disease, CTD [86] (http://www.biograph.be). Gene2MeSH: automatically annotates genes with the concepts defined in MeSH from PubMed abstracts [84] (http://gene2mesh.ncibi.org). Coremine: co-occurrence of terms from these lists creates relationship networks; the text records indexed are those comprising the MEDLINE data base (http://www.coremine.com/medical/#search). SNPs3d: abstracts of all the MEDLINE entries associated with each gene in the NCBI Gene Database [56] are the source of words and terms [85] (http://www.snps3d.org). neXtProt: integrates heterogeneous knowledge bases, i.e. IntAct, GO, Swiss-Prot [87] (http://www.nextprot.org). CIPA = Congenital insensitivity to pain with anhidrosis.

different text mining tools differ. For example, in a combined list containing putative genes involved in both depression and pain produced by the web tools mentioned above, only 5% of the genes were detected by two tools, 2% by three tools, and no gene was detected simultaneously by more than three tools (tables 2, 3).

We propose that different gene-prioritizing tools should be considered complementary rather than mutually exclusive. Importantly, any gene candidates produced by such tools should be considered as putative and should be verified. Efficiency of verification of the candidates revealed by data mining tools can also be greatly increased by utilizing web resources. Dedicated databases containing lists of depression- and pain-related genes are freely available online. Multi-level Knowledge base

Table 2. List of validated genes [present in manually curated databases of pain (PGdb)- and depression (MK4MDD)-related genes] involved in both depression and pain

Gene name	Data mining tool					Curated databases	
	BioGraph	Gene2MeSh	Coremine	neXtProt	SNPs3d	PGdb[1]	MK4MDD
BDNF	x	ND	x	ND	ND	x	x
CNR2	x	ND	x	x	ND	x	x
COMT	ND	ND	x	x	x	x	x
CYP19A1	x	ND	x	ND	ND	x	x
DRD2	ND	ND	x	x	x	x	x
ESR1	ND	ND	x	ND	x	x	x
GRIN2B	ND	ND	x	x	ND	x	x
HTR1A	ND	ND	x	x	ND	x	x
HTR1B	ND	ND	x	x	ND	x	x
HTR2A	ND	ND	x	x	x	x	x
KCNK2	ND	ND	x	x	ND	x	x
OPRM1	x	ND	x	ND	x	x	x
SLC6A2	ND	ND	x	ND	x	x	x
SLC6A4	ND	ND	x	x	x	x	x
TAC1	ND	ND	x	ND	x	x	x
TNF	ND	ND	x	x	x	x	x

ND = Not detected. [1] Validation based on mouse genes.

for Major Depressive Disorder (MK4MDD; http://mdd.psych.ac.cn) [66] is a manually curated and annually updated database containing various information on depression, including genes studied in the context of this disease. PGdb [28] is a manually curated and weekly updated database containing information on pain-related phenotypes of mice with gene knockouts.

Here, we present an example of the workflow used to select and validate a list of genes related to depression and pain. Table 1 presents queries used to retrieve lists of candidate genes for both depression and pain from the Gene2MeSH, SNPs3d, Coremine, BioGraph and neXtProt tools. After retrieving and combining the lists, we validated whether the putative candidate genes are indeed connected to depression. For this purpose, we compared the list of the candidate genes to the list containing all of the genes included in the MK4MDD database. Validation of involvement of candidate genes in pain was less straightforward, as the dataset of pain-related genes from PGdb is not available for downloading. Hence, each candidate gene name had to be manually typed into PGdb. Therefore, we used PGdb to verify only the most likely candidate genes (genes that were detected by two or three data mining tools or detected by one data mining tool and also present in MD4MDD database). Genes were considered to be involved in depression or pain according to information provided by the appropriate database (MD4MDD: ratio of number of studies reporting involvement of a given gene in depression to number of studies reporting lack of involvement had to be higher than 1; PGdb: directionality of change in pain-related traits after gene knockout had

Table 3. List of selected candidate (not validated manually, but detected by at least 2 different data mining tools) genes (either human or nonhuman) involved in both depression and pain

Gene name	Data mining tools					Curated databases	
	BioGraph	Gene2MeSh	Coremine	neXtProt	SNPs3d	PGdb	MK4MDD
ACE	ND	ND	x	ND	x	ND	x
CHRNA4	x	ND	x	x	ND	x	ND
CNR1	x	ND	x	ND	x	x	x
CRHR1	ND	ND	x	ND	x	ND	x
CRP	ND	ND	x	x	ND	ND	x
CYP2D6	ND	ND	x	x	x	ND	ND
FKBP5	ND	ND	x	x	ND	ND	x
GAL	ND	ND	x	ND	x	x	x
GNB3	ND	ND	ND	x	x	ND	x
HTR2C	x	ND	x	x	ND	ND	x
ICAM1	ND	ND	x	ND	x	ND	ND
IDO1	ND	ND	x	x	ND	x	ND
IGF1	ND	ND	x	x	ND	ND	ND
IL10	ND	ND	x	x	ND	x	x
IL1A	ND	ND	x	x	ND	ND	x
IL1B	ND	ND	ND	x	x	ND	x
MAOA	ND	ND	x	x	ND	ND	x
NTRK2	ND	ND	x	x	ND	ND	x
OPRL1	ND	ND	x	ND	x	x	ND
OXT	ND	ND	x	x	ND	x	ND
PDYN	ND	ND	x	x	ND	x	ND
PNOC	ND	ND	x	ND	x	x	ND
POMC	ND	ND	x	x	x	x	ND
PRL	ND	ND	x	x	ND	x	ND
SLC6A3	ND	ND	x	x	ND	ND	x
SOD1	ND	ND	x	x	ND	ND	ND
TACR1	ND	ND	x	ND	x	x	ND

ND = Not detected.

to be consistent). In this way we rapidly produced a list (table 2) of 16 validated candidate genes for a depression × pain interaction. Moreover, table 3 shows 27 genes that were detected by at least two different data mining tools but did not produce hits in either of the databases as potential candidate genes for depression × pain interaction. Validation of these genes goes beyond the scope of this short review, but it should be considered in future experimental studies. Additionally, the list of the validated candidates was enriched (associated) with overrepresented (compared with list of all the genes of the organism) biological processes and a biochemical pathway using the WebGestalt tool [88] (http://bioinfo.vanderbilt.edu/webgestalt/; table 4). This allows identification of putative higher-level (such as physiological) processes that participate in both depression and pain neuropathology. All together, the listed genes may be common for pain and depression disorders. Furthermore, expression of at least some of these genes may be epigenetically regulated.

Table 4. Selected annotations (categories of biological entities or terms[1]) enriched in validated candidate genes[2] involved in depression and pain (listed in table 2)

Database	Categories of biological entities (terms)	Validated genes involved in pain and depression
GO: biological process	Regulation of transmission of nerve impulse	*HTR1B, TNF, BDNF, CNR2, DRD2, SLC6A4, TAC1, HTR2A, GRIN2B*
GO: biological process	Response to drug	*HTR1B, TNF, COMT, BDNF, CYP19A1, DRD2, SLC6A4, HTR2A*
GO: biological process	Sensory perception of pain	*CNR2, TAC1, OPRM1, HTR2A, GRIN2B*
GO: biological process	Memory	*TAC1, DRD2, SLC6A4, HTR2A, GRIN2B*
GO: biological process	Regulation of synaptic transmission, GABAergic	*HTR1B, CNR2, TAC1, DRD2*
GO: biological process	Response to morphine	*CYP19A1, TAC1, OPRM1, DRD2*
GLAD4U: disease	Eating disorders	*HTR1A, HTR1B, BDNF, CNR2, DRD2, SLC6A4, COMT, TAC1, OPRM1, HTR2A, SLC6A2, GRIN2B*
GLAD4U: disease	Alcoholism	*HTR1A, HTR1B, BDNF, CNR2, DRD2, SLC6A4, COMT, OPRM1, HTR2A, SLC6A2, GRIN2B*
GLAD4U: disease	Headache	*HTR1A, HTR1B, TNF, BDNF, DRD2, SLC6A4, COMT, TAC1, HTR2A*
GLAD4U: disease	Depression	*HTR1A, HTR1B, BDNF, DRD2, SLC6A4, KCNK2, COMT, OPRM1, HTR2A, SLC6A2, GRIN2B*
GLAD4U: disease	Anxiety disorders	*HTR1A, HTR1B, BDNF, DRD2, SLC6A4, COMT, TAC1, HTR2A, SLC6A2, GRIN2B*
GLAD4U: disease	Opioid-related disorders	*HTR1B, COMT, BDNF, OPRM1, DRD2, SLC6A4, HTR2A*
GLAD4U: disease	Pain	*TNF, KCNK2, COMT, BDNF, TAC1, OPRM1, SLC6A4, HTR2A*
GLAD4U: disease	Schizophrenia	*HTR1A, HTR1B, COMT, BDNF, DRD2, SLC6A4, HTR2A, SLC6A2, GRIN2B*
GLAD4U: drug	Fluoxetine	*HTR1A, HTR1B, KCNK2, COMT, BDNF, SLC6A4, HTR2A, SLC6A2*
GLAD4U: drug	Dopamine	*HTR1A, HTR1B, BDNF, DRD2, SLC6A4, COMT, OPRM1, HTR2A, SLC6A2, GRIN2B*
GLAD4U: drug	Norepinephrine	*HTR1A, HTR1B, COMT, BDNF, DRD2, SLC6A4, HTR2A, SLC6A2*
GLAD4U: drug	Amphetamine	*COMT, BDNF, OPRM1, DRD2, SLC6A4, HTR2A, SLC6A2*

[1] Complete chains of events or molecular functions leading to a phenotype. [2] Genes identified in human and animal (mainly mouse) studies and analysis based on human databases and processes identified in *Homo sapiens*.

Epigenetic Mechanisms

Epigenetics encompasses all of the molecular mechanisms that stably alter gene expression without changing the DNA sequence and have potential to be transmitted through mitotic cell division. These mechanisms work mostly via regulation of the condensation state of chromatin (heterochromatin vs. euchromatin). Several interconnected epigenetic mechanisms have been identified, the most frequently studied of which are DNA methylation and histone modifications. Basic mechanisms relevant to neurobiological problems have been described in sufficient detail in several reviews [89–91].

DNA methylation relies on addition of methyl groups to cytosines. Methylation of cytosines is usually located in genomic loci enriched in CG dinucleotides, so-called CpG (cytosine-phosphate-guanidine) islands or CpG island shores [92, 93]. DNA methylation at the gene promoter is well recognized and generally, but not always [94], results in gene silencing. Histone modifications rely on addition of various chemical groups to the free histone tail. These chemical groups include methyl, acetyl and phosphate groups. The type of histone modification generally determines its effect on gene expression, but it was suggested that in some organisms (yeast), even a single type of histone modification may cause opposite transcriptional effects depending on the chromatin context [95, 96]. Therefore, epigenetic studies are complex and much less straightforward than studies of proteins or transcripts.

Epigenetics of Depression

Interest in epigenetics of depression has increased dramatically since 2012. Genome-wide studies using human peripheral DNA sources, such as blood or saliva, have uncovered a number of CpG sites whose differential methylation has been connected to depression. Some DNA methylation studies were performed on postmortem brains (mostly prefrontal cortex) of depressive patients. The results revealed changes in transcript levels of DNA methyltransferase (the protein responsible for setting and rewriting DNA methylation patterns) [97], as well as differential methylation of several genomic loci connected with genes such as *PRIMA1* (proline-rich membrane anchor 1) [98], *GABRA1* [GABA(A) receptor subunit alpha-1] [97] or *STK32C* (serine/threonine kinase 32C) [99]. Moreover, it was found that multiple classes of antidepressants produce changes in epigenetic mechanisms in brain cells derived from animal models or in human cell lines [100–103]. Similarly, substances that directly influence epigenetic patterns have been shown to reverse epigenetic changes and depression-like behavior induced in animal behavioral models of depression [104–107]. Hopefully, these compounds which act as inhibitors of DNA methyltransferases or, in particular, histone deacetylases (regulators of epigenetic histone tail acetylation) will be developed into novel therapies for depression, especially in cooperation with classical antidepressants [108]. As for now, four human genes have been of utmost interest to epigenetic-depression studies: *BDNF* (brain-derived neurotrophic factor) [109–111], *MAOA* (monoamine oxidase A) [112, 113], *NR3C1* (glucocorticoid receptor) [112, 114] and *SLC6A4* (serotonin transporter) [115–117]. All but *NR3C1* are listed in tables 2 or 3. All four genes encode proteins that have been strongly associated with depression [118–122].

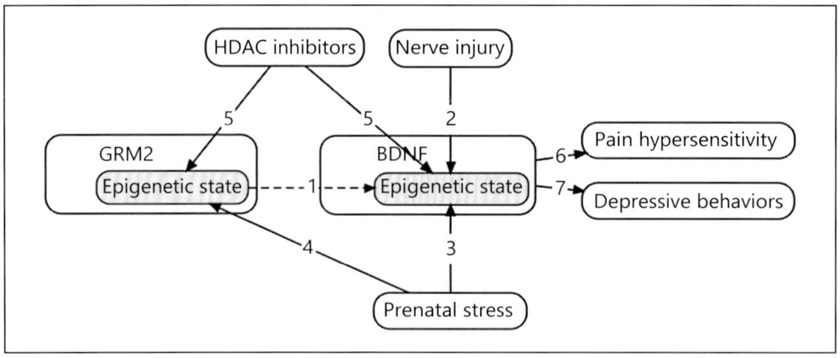

Fig. 1. Schematic of exemplary interactions between the selected, described genetic and epigenetic factors that may underlie pain × depression comorbidity – based on the common *GRM2* (glutamate receptor, metabotropic 2)-*BDNF* pathway in pain and depression.

Epigenetics of Pain

Epigenetics is also widely studied in pain. For example, several types of histone deacetylase inhibitors have displayed analgesic effects in various studies [123–125]. Animal studies have provided evidence that chronic stress [126], chronic inflammatory pain [127] and nerve injury [128] result in changes in epigenetic mechanisms that mediate pain sensitization. This pain sensitization can in turn be reversed by substances that target the underlying epigenetic changes [126–129]. Pain-related epigenetic changes have been observed both in peripheral tissues, such as immune cells gathering in the place of injury [130] or tumors [131], and in the central nervous system, i.e. in the nucleus raphe magnus [132] or prefrontal cortex [133]. There are also data suggesting that similar mechanisms may function in humans: whole-blood genome-wide DNA methylation has indicated that the pain gene *TRPA1* is linked with pain sensitivity [134]. Moreover, epigenetic mechanisms seem to participate in antinociceptive drug activity. Chronic opioid intake causes hyperalgesia and this effect has been shown, in mice, to be dependent on decreased histone deacetylase activity and increased expression of *Bdnf* in the spinal cord [135]. In humans, chronic opioid use is associated with increased global DNA methylation, including increased methylation of the *OPRM1* gene coding for the mu-opioid receptor, and correlated with clinical pain [136].

Epigenetics of Pain and Depression

It is clear that epigenetic mechanisms may participate in both pain and depression. For example, as indicated above, histone deacetylase (HDAC) inhibitors have already been considered as therapeutic agents for both depression and pain [137]. Furthermore, there have been some attempts to assign (common) epigenetic mechanisms to pain and mental health disorders [19, 138]. It can be proposed that some aspects of pain and depression may be regulated by the same genes and epigenetic mechanisms (fig. 1):

(1) Epigenetic induction of *Grm2* expression results in antidepressive effects in rats and mice [107]; in the mouse brain, *Grm2* regulates DNA methylation [139]; *Grm2* mediates antidepressive effects in rats via modulation of *Bdnf* expression [140], and in the spinal cord, *Grm2* mediates antinociceptive effect of HDAC inhibitors in rats and mice [141, 142]
(2) Sciatic nerve injury causes depressive-like behavior in mice [143]; in the spinal cord, *Bdnf* is epigenetically upregulated after sciatic nerve injury and its expression mediates pain sensitization and neuropathic pain in mice [144]
(3) Chronic prenatal stress causes pain sensitization via epigenetic upregulation of *Bdnf* in the spinal cord [145], prenatal stress is an animal model of depression [146, 147] and prenatal stress and *BDNF* polymorphisms have been suggested to influence internalizing symptoms in children [148]
(4) Epigenetic changes of *Grm2* caused by prenatal stress induce a schizophrenia-like phenotype including a deficit in social interaction in mice [149]
(5) HDAC inhibitors alter epigenetic status (histone modifications) of *Grm2* [150] and *Bdnf* [151]; HDAC inhibitors attenuate depressive symptoms [104] and pain sensitivity in rats [125]
(6) *Bdnf* expression mediates pain sensitization [144, 145]
(7) *BDNF* gene constitutes a crucial element of depression pathophysiology and antidepressant response [122]

The hypothetical model above also suggests a role for epigenetics in connecting environment and genotype to phenotype and disease. Although no obvious epigenetic connection between depression and pain is currently known, it seems likely that such specific links will be revealed in the future.

Conclusions

Genotype contributes to nociception and psychiatric disorders. Psychiatric and pain disorder comorbidity could be mediated by common genes or entire functional pathways. The gene expression resulting in disease phenotype can be modified by epigenetic factors and mechanisms, pharmacotherapy, and drug candidates. Some of the factors (sex, nutrition, stress, etc.) appear to be common and implicated in the etiology of pain and psychiatric disorders. Also, the effects of environmental factors, both external and internal, on genotype are likely to be mediated by the epigenetic mechanisms.

Epigenetic mechanisms (including their dysregulation) may be involved in the two disorders and their comorbidity. Only a few genes have been definitively implicated in both disorders. However, it can be hypothesized that adverse life events through epigenetic modification elicit changes in expression of these particular genes, thus leading to both disorders. The possible contribution of epigenetic regulation to pain and psychiatric disorders suggests that treatments targeting epigenetic mechanisms should be considered.

References

1 Bair MJ, Robinson RL, Katon W, Kroenke K: Depression and pain comorbidity: a literature review. Arch Intern Med 2003;163:2433–2445.

2 Li JX: Pain and depression comorbidity: a preclinical perspective. Behav Brain Res 2015;276:92–98.

3 Blackburn-Munro G, Blackburn-Munro RE: Chronic pain, chronic stress and depression: coincidence or consequence? J Neuroendocrinol 2001;13:1009–1023.

4 Mogil JS: Pain genetics: past, present and future. Trends Genet 2012;28:258–266.

5 Mogil JS, Sternberg WF, Marek P, Sadowski B, Belknap JK, Liebeskind JC: The genetics of pain and pain inhibition. Proc Natl Acad Sci U S A 1996;93: 3048–3055.

6 Foulkes T, Wood JN: Pain genes. PLoS Genet 2008; 4:e1000086.

7 Lohoff FW: Overview of the genetics of major depressive disorder. Curr Psychiatry Rep 2010;12:539–546.

8 Sullivan PF, Daly MJ, O'Donovan M: Genetic architectures of psychiatric disorders: the emerging picture and its implications. Nat Rev Genet 2012;13: 537–551.

9 Flint J, Kendler KS: The genetics of major depression. Neuron 2014;81:484–503.

10 Cohen-Woods S, Craig IW, McGuffin P: The current state of play on the molecular genetics of depression. Psychol Med 2013;43:673–687.

11 Max MB, Wu T, Atlas SJ, Edwards RR, Haythornthwaite JA, Bollettino AF, Belfer I: A clinical genetic method to identify mechanisms by which pain causes depression and anxiety. Mol Pain 2006;2:14.

12 Maletic V, Raison CL: Neurobiology of depression, fibromyalgia and neuropathic pain. Front Biosci (Landmark Ed) 2009;14:5291–5338.

13 Denk F, McMahon SB: Chronic pain: emerging evidence for the involvement of epigenetics. Neuron 2012;73:435–444.

14 Stone LS, Szyf M: The emerging field of pain epigenetics. Pain 2013;154:1–2.

15 Crow M, Denk F, McMahon SB: Genes and epigenetic processes as prospective pain targets. Genome Med 2013;5:12.

16 Menke A, Binder EB: Epigenetic alterations in depression and antidepressant treatment. Dialogues Clin Neurosci 2014;16:395–404.

17 Pena CJ, Bagot RC, Labonte B, Nestler EJ: Epigenetic signaling in psychiatric disorders. J Mol Biol 2014; 426:3389–3412.

18 Dalton VS, Kolshus E, McLoughlin DM: Epigenetics and depression: return of the repressed. J Affect Disord 2014;155:1–12.

19 Rahn EJ, Guzman-Karlsson MC, David Sweatt J: Cellular, molecular, and epigenetic mechanisms in non-associative conditioning: implications for pain and memory. Neurobiol Learn Mem 2013;105:133–150.

20 Amanzio M, Pollo A, Maggi G, Benedetti F: Response variability to analgesics: a role for non-specific activation of endogenous opioids. Pain 2001;90: 205–215.

21 Nielsen CS, Price DD, Vassend O, Stubhaug A, Harris JR: Characterizing individual differences in heat-pain sensitivity. Pain 2005;119:65–74.

22 Panocka I, Marek P, Sadowski B: Inheritance of stress-induced analgesia in mice. Selective breeding study. Brain Res 1986;397:152–155.

23 Panocka I, Marek P, Sadowski B: Differentiation of neurochemical basis of stress-induced analgesia in mice by selective breeding. Brain Res 1986;397:156–160.

24 Mogil JS, Marek P, Flodman P, Spence MA, Sternberg WF, Kest B, Liebeskind JC: One or two genetic loci mediate high opiate analgesia in selectively bred mice. Pain 1995;60:125–135.

25 Lacroix-Fralish ML, Mogil JS: Progress in genetic studies of pain and analgesia. Annu Rev Pharmacol Toxicol 2009;49:97–121.

26 Walter C, Lotsch J: Meta-analysis of the relevance of the OPRM1 118A>G genetic variant for pain treatment. Pain 2009;146:270–275.

27 Lotsch J, Doehring A, Mogil JS, Arndt T, Geisslinger G, Ultsch A: Functional genomics of pain in analgesic drug development and therapy. Pharmacol Ther 2013;139:60–70.

28 Lacroix-Fralish ML, Ledoux JB, Mogil JS: The Pain Genes Database: an interactive web browser of pain-related transgenic knockout studies. Pain 2007;131: 3.e1–4.

29 Patel A: Review: the role of inflammation in depression. Psychiatr Danub 2013;25(suppl 2):S216–S223.

30 Dunn AJ, Swiergiel AH, de Beaurepaire R: Cytokines as mediators of depression: what can we learn from animal studies? Neurosci Biobehav Rev 2005;29: 891–909.

31 Illi J, Miaskowski C, Cooper B, Levine JD, Dunn L, West C, Aouizerat BE: Association between pro- and anti-inflammatory cytokine genes and a symptom cluster of pain, fatigue, sleep disturbance, and depression. Cytokine 2012;58:437–447.

32 Reyes-Gibby CC, Wang J, Spitz M, Wu X, Yennurajalingam S, Shete S: Genetic variations in interleukin-8 and interleukin-10 are associated with pain, depressed mood, and fatigue in lung cancer patients. J Pain Symptom Manag 2013;46:161–172.

33 Holliday KL, Nicholl BI, Macfarlane GJ, Thomson W, Davies KA, McBeth J: Genetic variation in the hypothalamic-pituitary-adrenal stress axis influences susceptibility to musculoskeletal pain: results from the EPIFUND study. Ann Rheum Dis 2010;69: 556–560.
34 Holliday KL, Macfarlane GJ, Nicholl BI, Creed F, Thomson W, McBeth J: Genetic variation in neuroendocrine genes associates with somatic symptoms in the general population: results from the EPIFUND study. J Psychosom Res 2010;68:469–474.
35 Darnaudery M, Maccari S: Epigenetic programming of the stress response in male and female rats by prenatal restraint stress. Brain Res Rev 2008;57:571–585.
36 Marek P, Panocka I, Sadowski B: Involvement of humoral factors in the mechanism of stress-induced analgesia in mice. Ann N Y Acad Sci 1986;467:361–370.
37 Butler RK, Finn DP: Stress-induced analgesia. Prog Neurobiol 2009;88:184–202.
38 Jennings EM, Okine BN, Roche M, Finn DP: Stress-induced hyperalgesia. Prog Neurobiol 2014;121:1–18.
39 Olango WM, Finn DP: Neurobiology of stress-induced hyperalgesia. Curr Top Behav Neurosci 2014; 20:251–280.
40 Zorumski CF, Paul SM, Izumi Y, Covey DF, Mennerick S: Neurosteroids, stress and depression: potential therapeutic opportunities. Neurosci Biobehav Rev 2013;37:109–122.
41 Pittenger C, Duman RS: Stress, depression, and neuroplasticity: a convergence of mechanisms. Neuropsychopharmacology 2008;33:88–109.
42 aan het Rot M, Mathew SJ, Charney DS: Neurobiological mechanisms in major depressive disorder. CMAJ 2009;180:305–313.
43 Lacroix-Fralish ML: Sex-specific pain modulation: the growth factor, neuregulin-1, as a pro-nociceptive cytokine. Neurosci Lett 2008;437:184–187.
44 Wiesenfeld-Hallin Z: Sex differences in pain perception. Gend Med 2005;2:137–145.
45 Myers CD, Riley JL, 3rd, Robinson ME: Psychosocial contributions to sex-correlated differences in pain. Clin J Pain 2003;19:225–232.
46 Fillingim RB: Sex, gender, and pain: women and men really are different. Curr Rev Pain 2000;4:24–30.
47 Altemus M, Sarvaiya N, Neill Epperson C: Sex differences in anxiety and depression clinical perspectives. Front Neuroendocrinol 2014;35:320–330.
48 Seney ML, Sibille E: Sex differences in mood disorders: perspectives from humans and rodent models. Biol Sex Diff 2014;5:17.
49 Bair MJ, Robinson RL, Katon W, Kroenke K: Depression and pain comorbidity: a literature review. Arch Intern Med 2003;163:2433–2445.
50 Meana M: The meeting of pain and depression: comorbidity in women. Can J Psychiatry 1998;43:893–899.
51 Mogil JS, Sorge RE, LaCroix-Fralish ML, Smith SB, Fortin A, Sotocinal SG, Fillingim RB: Pain sensitivity and vasopressin analgesia are mediated by a gene-sex-environment interaction. Nat Neurosci 2011;14: 1569–1573.
52 Mogil JS, Richards SP, O'Toole LA, Helms ML, Mitchell SR, Kest B, Belknap JK: Identification of a sex-specific quantitative trait locus mediating nonopioid stress-induced analgesia in female mice. J Neurosci 1997;17:7995–8002.
53 Mogil JS, Belknap JK: Sex and genotype determine the selective activation of neurochemically-distinct mechanisms of swim stress-induced analgesia. Pharmacol Biochem Behav 1997;56:61–66.
54 Stankiewicz AM, Swiergiel AH, Lisowski P: Epigenetics of stress adaptations in the brain. Brain Res Bull 2013;98:76–92.
55 Taylor L, Faraone SV, Tsuang MT: Family, twin, and adoption studies of bipolar disease. Curr Psychiatry Rep 2002;4:130–133.
56 Lee SH, Ripke S, Neale BM, Faraone SV, Purcell SM, Perlis RH, Wray NR: Genetic relationship between five psychiatric disorders estimated from genome-wide SNPs. Nat Genet 2013;45:984–994.
57 Bosker FJ, Hartman CA, Nolte IM, Prins BP, Terpstra P, Posthuma D, Nolen WA: Poor replication of candidate genes for major depressive disorder using genome-wide association data. Mol Psychiatry 2011; 16:516–532.
58 Caspi A, Hariri AR, Holmes A, Uher R, Moffitt TE: Genetic sensitivity to the environment: the case of the serotonin transporter gene and its implications for studying complex diseases and traits. Am J Psychiatry 2010;167:509–527.
59 Wendland JR, Martin BJ, Kruse MR, Lesch KP, Murphy DL: Simultaneous genotyping of four functional loci of human SLC6A4, with a reappraisal of 5-HTTLPR and rs25531. Mol Psychiatry 2006;11: 224–226.
60 Alexander N, Klucken T, Koppe G, Osinsky R, Walter B, Vaitl D, Hennig J: Interaction of the serotonin transporter-linked polymorphic region and environmental adversity: increased amygdala-hypothalamus connectivity as a potential mechanism linking neural and endocrine hyperreactivity. Biol Psychiatry 2012;72:49–56.
61 Wankerl M, Wust S, Otte C: Current developments and controversies: does the serotonin transporter gene-linked polymorphic region (5-HTTLPR) modulate the association between stress and depression? Curr Opin Psychiatry 2010;23:582–587.

62 Kinnally EL, Capitanio JP, Leibel R, Deng L, LeDuc C, Haghighi F, Mann JJ: Epigenetic regulation of serotonin transporter expression and behavior in infant rhesus macaques. Genes Brain Behav 2010;9: 575–582.
63 Jutkiewicz EM: The antidepressant-like effects of delta-opioid receptor agonists. Mol Interv 2006;6: 162–169.
64 Jutkiewicz EM, Roques BP: Endogenous opioids as physiological antidepressants: complementary role of delta receptors and dopamine. Neuropsychopharmacology 2012;37:303–304.
65 Lutz PE, Kieffer BL: Opioid receptors: distinct roles in mood disorders. Trends Neurosci 2013;36:195–206.
66 Guo L, Zhang W, Chang S, Zhang L, Ott J, Wang J: MK4MDD: a multi-level knowledge base and analysis platform for major depressive disorder. PLoS One 2012;7:e46335.
67 Jann MW, Slade JH: Antidepressant agents for the treatment of chronic pain and depression. Pharmacotherapy 2007;27:1571–1587.
68 Lebe M, Hasenbring MI, Schmieder K, Jetschke K, Harders A, Epplen JT, Kotting J: Association of serotonin-1A and -2A receptor promoter polymorphisms with depressive symptoms, functional recovery, and pain in patients 6 months after lumbar disc surgery. Pain 2013;154:377–384.
69 Sadowski B, Panocka I: Cross-tolerance between morphine and swim analgesia in mice selectively bred for high and low stress-induced analgesia. Pharmacol Biochem Behav 1993;45:527–531.
70 Lutfy K, Sadowski B, Kwon IS, Weber E: Morphine analgesia and tolerance in mice selectively bred for divergent swim stress-induced analgesia. Eur J Pharmacol 1994;265:171–174.
71 Juszczak GR, Sliwa AT, Wolak P, Tymosiak-Zielinska A, Lisowski P, Swiergiel AH: The usage of video analysis system for detection of immobility in the tail suspension test in mice. Pharmacol Biochem Behavior 2006;85:332–338.
72 Lutz PE, Kieffer BL: Opioid receptors: distinct roles in mood disorders. Trends Neurosci 2013;36:195–206.
73 Panocka I, Massi M, Lapo I, Swiderski T, Kowalczyk M, Sadowski B: Antidepressant-type effect of the NK3 tachykinin receptor agonist aminosenktide in mouse lines differing in endogenous opioid system activity. Peptides 2001;22:1037–1042.
74 Juszczak GR, Lisowski P, Sliwa AT, Swiergiel AH: Computer assisted video analysis of swimming performance in a forced swim test: simultaneous assessment of duration of immobility and swimming style in mice selected for high and low swim-stress induced analgesia. Physiol Behav 2008;95:400–407.
75 Lisowski P, Juszczak GR, Goscik J, Stankiewicz AM, Wieczorek M, Zwierzchowski L, Swiergiel AH: Stress susceptibility-specific phenotype associated with different hippocampal transcriptomic responses to chronic tricyclic antidepressant treatment in mice. BMC Neurosci 2013;14:144.
76 Lisowski P, Juszczak GR, Goscik J, Wieczorek M, Zwierzchowski L, Swiergiel AH: Effect of chronic mild stress on hippocampal transcriptome in mice selected for high and low stress-induced analgesia and displaying different emotional behaviors. Eur Neuropsychopharmacol 2011;21:45–62.
77 Li JX: Pain and depression comorbidity: a preclinical perspective. Behav Brain Res 2015;276:92–98.
78 Nicholl BI, Holliday KL, Macfarlane GJ, Thomson W, Davies KA, O'Neill TW, McBeth J: Association of HTR2A polymorphisms with chronic widespread pain and the extent of musculoskeletal pain: results from two population-based cohorts. Arthritis Rheum 2011;63:810–818.
79 Lindstedt F, Lonsdorf TB, Schalling M, Kosek E, Ingvar M: Perception of thermal pain and the thermal grill illusion is associated with polymorphisms in the serotonin transporter gene. PLoS One 2011;6: e17752.
80 Fijal B, Perlis RH, Heinloth AN, Houston JP: The association of single nucleotide polymorphisms in the catechol-O-methyltransferase gene and pain scores in female patients with major depressive disorder. J Pain 2010;11:910–915, 915.e1–9.
81 Ablin JN, Bar-Shira A, Yaron M, Orr-Urtreger A: Candidate-gene approach in fibromyalgia syndrome: association analysis of the genes encoding substance P receptor, dopamine transporter and alpha1-antitrypsin. Clin Exp Rheumatol 2009;27:S33–S38.
82 Burke NN, Geoghegan E, Kerr DM, Moriarty O, Finn DP, Roche M: Altered neuropathic pain behaviour in a rat model of depression is associated with changes in inflammatory gene expression in the amygdala. Genes Brain Behav 2013;12:705–713.
83 Mirkovic K, Palmersheim J, Lesage F, Wickman K: Behavioral characterization of mice lacking Trek channels. Front Behav Neurosci 2012;6:60.
84 Ade AS, Wright ZC, States DJ: Gene2MeSH (Internet). Ann Arbor, National Center for Integrative Biomedical Informatics, 2007.
85 Yue P, Melamud E, Moult J: SNPs3D: candidate gene and SNP selection for association studies. BMC Bioinformatics 2006;7:166.
86 Liekens AM, De Knijf J, Daelemans W, Goethals B, De Rijk P, Del-Favero J: BioGraph: unsupervised biomedical knowledge discovery via automated hypothesis generation. Genome Biol 2011;12:R57.

87 Lane L, Argoud-Puy G, Britan A, Cusin I, Duek PD, Evalet O, Bairoch A: neXtProt: a knowledge platform for human proteins. Nucleic Acids Res 2012;40:D76–D83.
88 Wang J, Duncan D, Shi Z, Zhang B: WEB-based GEne SeT AnaLysis Toolkit (WebGestalt): update 2013. Nucleic Acids Res 2013;41:W77–W83.
89 Chouliaras L, Rutten BP, Kenis G, Peerbooms O, Visser PJ, Verhey F, van den Hove DL: Epigenetic regulation in the pathophysiology of Alzheimer's disease. Prog Neurobiol 2010;90:498–510.
90 Robison AJ, Nestler EJ: Transcriptional and epigenetic mechanisms of addiction. Nat Rev Neurosci 2011;12:623–637.
91 Buchheit T, Van de Ven T, Shaw A: Epigenetics and the transition from acute to chronic pain. Pain Med 2012;13:1474–1490.
92 Haines TR, Rodenhiser DI, Ainsworth PJ: Allele-specific non-CpG methylation of the Nf1 gene during early mouse development. Dev Biol 2001;240: 585–598.
93 Guo JU, Su Y, Shin JH, Shin J, Li H, Xie B, Song H: Distribution, recognition and regulation of non-CpG methylation in the adult mammalian brain. Nat Neurosci 2014;17:215–222.
94 Bahar Halpern K, Vana T, Walker MD: Paradoxical role of DNA methylation in activation of FoxA2 gene expression during endoderm development. J Biol Chem 2014;289:23882–23892.
95 Pinskaya M, Morillon A: Histone H3 lysine 4 dimethylation: a novel mark for transcriptional fidelity? Epigenetics 2009;4:302–306.
96 Rando OJ: Combinatorial complexity in chromatin structure and function: revisiting the histone code. Curr Opin Genet Dev 2012;22:148–155.
97 Poulter MO, Du L, Weaver IC, Palkovits M, Faludi G, Merali Z, Anisman H: GABAA receptor promoter hypermethylation in suicide brain: implications for the involvement of epigenetic processes. Biol Psychiatry 2008;64:645–652.
98 Sabunciyan S, Aryee MJ, Irizarry RA, Rongione M, Webster MJ, Kaufman WE, Gen REDC: Genome-wide DNA methylation scan in major depressive disorder. PLoS One 2012;7:e34451.
99 Dempster EL, Wong CC, Lester KJ, Burrage J, Gregory AM, Mill J, Eley TC: Genome-wide methylomic analysis of monozygotic twins discordant for adolescent depression. Biol Psychiatry 2014;76:977–983.
100 Ookubo M, Kanai H, Aoki H, Yamada N: Antidepressants and mood stabilizers effects on histone deacetylase expression in C57BL/6 mice: brain region specific changes. J Psychiatr Res 2013;47: 1204–1214.
101 Lee MG, Wynder C, Schmidt DM, McCafferty DG, Shiekhattar R: Histone H3 lysine 4 demethylation is a target of nonselective antidepressive medications. Chem Biol 2006;13:563–567.
102 Tsankova NM, Berton O, Renthal W, Kumar A, Neve RL, Nestler EJ: Sustained hippocampal chromatin regulation in a mouse model of depression and antidepressant action. Nat Neurosci 2006;9: 519–525.
103 Robison AJ, Vialou V, Sun HS, Labonte B, S AG, Dias C, Nestler EJ: Fluoxetine epigenetically alters the CaMKIIα promoter in nucleus accumbens to regulate ΔFosB binding and antidepressant effects. Neuropsychopharmacology 2014;39:1178–1186.
104 Covington HE 3rd, Maze I, LaPlant QC, Vialou VF, Ohnishi YN, Berton O, Nestler EJ: Antidepressant actions of histone deacetylase inhibitors. J Neurosci 2009;29:11451–11460.
105 Zhu H, Huang Q, Xu H, Niu L, Zhou JN: Antidepressant-like effects of sodium butyrate in combination with estrogen in rat forced swimming test: involvement of 5-HT(1A) receptors. Behav Brain Res 2009;196:200–206.
106 Xing B, Liu P, Xu WJ, Xu FY, Dang YH: Effect of microinjecting of 5-aza-2-deoxycytidine into ventrolateral orbital cortex on depressive-like behavior in rats. Neurosci Lett 2014;574:11–14.
107 Nasca C, Xenos D, Barone Y, Caruso A, Scaccianoce S, Matrisciano F, Nicoletti F: L-acetylcarnitine causes rapid antidepressant effects through the epigenetic induction of mGlu2 receptors. Proc Natl Acad Sci U S A 2013;110:4804–4809.
108 Grayson DR, Kundakovic M, Sharma RP: Is there a future for histone deacetylase inhibitors in the pharmacotherapy of psychiatric disorders? Mol Pharmacol 2010;77:126–135.
109 Dell'Osso B, D'Addario C, Carlotta Palazzo M, Benatti B, Camuri G, Galimberti D, Altamura AC: Epigenetic modulation of BDNF gene: differences in DNA methylation between unipolar and bipolar patients. J Affect Disord 2014;166:330–333.
110 Fuchikami M, Morinobu S, Segawa M, Okamoto Y, Yamawaki S, Ozaki N, Terao T: DNA methylation profiles of the brain-derived neurotrophic factor (BDNF) gene as a potent diagnostic biomarker in major depression. PLoS One 2011;6:e23881.
111 Carlberg L, Scheibelreiter J, Hassler MR, Schloegelhofer M, Schmoeger M, Ludwig B, Schosser A: Brain-derived neurotrophic factor (BDNF)-epigenetic regulation in unipolar and bipolar affective disorder. J Affect Disord 2014;168:399–406.
112 Melas PA, Wei Y, Wong CC, Sjoholm LK, Aberg E, Mill J, Lavebratt C: Genetic and epigenetic associations of MAOA and NR3C1 with depression and childhood adversities. Int J Neuropsychopharmacol 2013;16:1513–1528.
113 Domschke K, Tidow N, Schwarte K, Ziegler C, Lesch KP, Deckert J, Baune BT: Pharmacoepigenetics of depression: no major influence of MAO-A DNA methylation on treatment response. J Neural Transm 2015;122:99–108.

114 Na KS, Chang HS, Won E, Han KM, Choi S, Tae WS, Ham BJ: Association between glucocorticoid receptor methylation and hippocampal subfields in major depressive disorder. PLoS One 2014;9: e85425.
115 Olsson CA, Foley DL, Parkinson-Bates M, Byrnes G, McKenzie M, Patton GC, Saffery R: Prospects for epigenetic research within cohort studies of psychological disorder: a pilot investigation of a peripheral cell marker of epigenetic risk for depression. Biol Psychol 2010;83:159–165.
116 Kang HJ, Kim JM, Stewart R, Kim SY, Bae KY, Kim SW, Yoon JS: Association of SLC6A4 methylation with early adversity, characteristics and outcomes in depression. Prog Neuropsychopharmacol Biol Psychiatry 2013;44:23–28.
117 Zhao J, Goldberg J, Bremner JD, Vaccarino V: Association between promoter methylation of serotonin transporter gene and depressive symptoms: a monozygotic twin study. Psychosom Med 2013;75: 523–529.
118 Karg K, Burmeister M, Shedden K, Sen S: The serotonin transporter promoter variant (5-HTTLPR), stress, and depression meta-analysis revisited: evidence of genetic moderation. Arch Gen Psychiatry 2011;68:444–454.
119 Szczepankiewicz A, Leszczynska-Rodziewicz A, Pawlak J, Rajewska-Rager A, Dmitrzak-Weglarz M, Wilkosc M, Hauser J: Glucocorticoid receptor polymorphism is associated with major depression and predominance of depression in the course of bipolar disorder. J Affect Disord 2011;134:138–144.
120 Dannlowski U, Ohrmann P, Konrad C, Domschke K, Bauer J, Kugel H, Suslow T: Reduced amygdala-prefrontal coupling in major depression: association with MAOA genotype and illness severity. Int J Neuropsychopharmacol 2009;12:11–22.
121 Aklillu E, Karlsson S, Zachrisson OO, Ozdemir V, Agren H: Association of MAOA gene functional promoter polymorphism with CSF dopamine turnover and atypical depression. Pharmacogenet Genomics 2009;19:267–275.
122 Castren E, Rantamaki T: The role of BDNF and its receptors in depression and antidepressant drug action: reactivation of developmental plasticity. Dev Neurobiol 2010;70:289–297.
123 Kukkar A, Singh N, Jaggi AS: Attenuation of neuropathic pain by sodium butyrate in an experimental model of chronic constriction injury in rats. J Formos Med Assoc 2013.
124 Yoshizumi M, Eisenach JC, Hayashida K: Valproate prevents dysregulation of spinal glutamate and reduces the development of hypersensitivity in rats after peripheral nerve injury. J Pain 2013;14:1485–1491.
125 Denk F, Huang W, Sidders B, Bithell A, Crow M, Grist J, McMahon SB: HDAC inhibitors attenuate the development of hypersensitivity in models of neuropathic pain. Pain 2013;154:1668–1679.
126 Hong S, Zheng G, Wiley JW: Epigenetic regulation of genes that modulate chronic stress-induced visceral pain in the peripheral nervous system. Gastroenterology 2015;148:148–157.e7.
127 Pan Z, Zhu LJ, Li YQ, Hao LY, Yin C, Yang JX, Cao JL: Epigenetic modification of spinal miR-219 expression regulates chronic inflammation pain by targeting CaMKIIγ. J Neurosci 2014;34:9476–9483.
128 Wang Y, Liu C, Guo QL, Yan JQ, Zhu XY, Huang CS, Zou WY: Intrathecal 5-azacytidine inhibits global DNA methylation and methyl-CpG-binding protein 2 expression and alleviates neuropathic pain in rats following chronic constriction injury. Brain Res 2011;1418:64–69.
129 Ueda H, Uchida H: Epigenetic modification in neuropathic pain. Curr Pharm Des 2015;21:849–867.
130 Kiguchi N, Kobayashi Y, Saika F, Kishioka S: Epigenetic upregulation of CCL2 and CCL3 via histone modifications in infiltrating macrophages after peripheral nerve injury. Cytokine 2013;64: 666–672.
131 Viet CT, Ye Y, Dang D, Lam DK, Achdjian S, Zhang J, Schmidt BL: Re-expression of the methylated EDNRB gene in oral squamous cell carcinoma attenuates cancer-induced pain. Pain 2011;152: 2323–2332.
132 Zhang Z, Cai YQ, Zou F, Bie B, Pan ZZ: Epigenetic suppression of GAD65 expression mediates persistent pain. Nat Med 2011;17:1448–1455.
133 Tajerian M, Alvarado S, Millecamps M, Vachon P, Crosby C, Bushnell MC, Stone LS: Peripheral nerve injury is associated with chronic, reversible changes in global DNA methylation in the mouse prefrontal cortex. PLoS One 2013;8:e55259.
134 Bell JT, Loomis AK, Butcher LM, Gao F, Zhang B, Hyde CL, Spector TD: Differential methylation of the TRPA1 promoter in pain sensitivity. Nat Commun 2014;5:2978.
135 Liang DY, Sun Y, Shi XY, Sahbaie P, Clark JD: Epigenetic regulation of spinal cord gene expression controls opioid-induced hyperalgesia. Mol Pain 2014;10:59.
136 Doehring A, Oertel BG, Sittl R, Lotsch J: Chronic opioid use is associated with increased DNA methylation correlating with increased clinical pain. Pain 2013;154:15–23.
137 Schroeder M, Krebs MO, Bleich S, Frieling H: Epigenetics and depression: current challenges and new therapeutic options. Curr Opin Psychiatry 2010;23:588–592.

138 Tran L, Schulkin J, Ligon CO, Greenwood-Van Meerveld B: Epigenetic modulation of chronic anxiety and pain by histone deacetylation. Mol Psychiatry 2014, Epub ahead of print.
139 Matrisciano F, Dong E, Gavin DP, Nicoletti F, Guidotti A: Activation of group II metabotropic glutamate receptors promotes DNA demethylation in the mouse brain. Mol Pharmacol 2011;80: 174–182.
140 Koike H, Fukumoto K, Iijima M, Chaki S: Role of BDNF/TrkB signaling in antidepressant-like effects of a group II metabotropic glutamate receptor antagonist in animal models of depression. Behav Brain Res 2013;238:48–52.
141 Cao DY, Bai G, Ji Y, Traub RJ: Epigenetic upregulation of metabotropic glutamate receptor 2 in the spinal cord attenuates oestrogen-induced visceral hypersensitivity. Gut 2014, Epub ahead of print.
142 Chiechio S, Zammataro M, Morales ME, Busceti CL, Drago F, Gereau RW 4th, Nicoletti F: Epigenetic modulation of mGlu2 receptors by histone deacetylase inhibitors in the treatment of inflammatory pain. Mol Pharmacol 2009;75:1014–1020.
143 Norman GJ, Karelina K, Zhang N, Walton JC, Morris JS, Devries AC: Stress and IL-1beta contribute to the development of depressive-like behavior following peripheral nerve injury. Mol Psychiatry 2010;15:404–414.
144 Uchida H, Matsushita Y, Ueda H: Epigenetic regulation of BDNF expression in the primary sensory neurons after peripheral nerve injury: implications in the development of neuropathic pain. Neuroscience 2013;240:147–154.
145 Winston JH, Li Q, Sarna SK: Chronic prenatal stress epigenetically modifies spinal cord BDNF expression to induce sex-specific visceral hypersensitivity in offspring. Neurogastroenterol Motil 2014; 26:715–730.
146 Mairesse J, Van Camp G, Gatta E, Marrocco J, Reynaert ML, Consolazione M, Maccari S: Sleep in prenatally restraint stressed rats, a model of mixed anxiety-depressive disorder. Adv Neurobiol 2015; 10:27–44.
147 Zhang XH, Jia N, Zhao XY, Tang GK, Guan LX, Wang D, Zhu ZL: Involvement of pGluR1, EAAT2 and EAAT3 in offspring depression induced by prenatal stress. Neuroscience 2013;250:333–341.
148 O'Donnell KJ, Glover V, Holbrook JD, O'Connor TG: Maternal prenatal anxiety and child brain-derived neurotrophic factor (BDNF) genotype: effects on internalizing symptoms from 4 to 15 years of age. Dev Psychopathol 2014;26:1255–1266.
149 Matrisciano F, Tueting P, Maccari S, Nicoletti F, Guidotti A: Pharmacological activation of group-II metabotropic glutamate receptors corrects a schizophrenia-like phenotype induced by prenatal stress in mice. Neuropsychopharmacology 2012; 37:929–938.
150 Kurita M, Holloway T, Garcia-Bea A, Kozlenkov A, Friedman AK, Moreno JL, Gonzalez-Maeso J: HDAC2 regulates atypical antipsychotic responses through the modulation of mGlu2 promoter activity. Nat Neurosci 2012;15:1245–1254.
151 Koppel I, Timmusk T: Differential regulation of Bdnf expression in cortical neurons by class-selective histone deacetylase inhibitors. Neuropharmacology 2013;75:106–115.

Prof. Artur H. Swiergiel
Department of Animal and Human Physiology
Faculty of Biology, Gdansk University
Wita Stwosza 59, PL–80-308 Gdansk (Poland)
E-Mail swiergiel@yahoo.com

Finn DP, Leonard BE (eds): Pain in Psychiatric Disorders. Mod Trends Pharmacopsychiatry.
Basel, Karger, 2015, vol 30, pp 138–152 (DOI: 10.1159/000435938)

Emotional and Cognitive Influences on Pain Experience

Madelon L. Peters

Department of Clinical Psychological Science, Maastricht University, Maastricht, The Netherlands

Abstract

Multiple emotional and cognitive factors impact on the experience of pain. This chapter will review some of the most important emotional and cognitive determinants of the pain experience as found in experimental and clinical studies with human participants. Emotional factors that may increase pain perception are anxiety, depression and anger. Positive emotions usually decrease perceived pain. The cognitive factors attention, expectancy and appraisal can either increase or decrease pain experiences depending on their specific focus and content. Many brain regions are involved in nociceptive processing and bringing pain into awareness. There are profound interconnections between areas processing sensory, emotional and cognitive information. Descending pathways from cortical areas to the midbrain and spinal levels can facilitate or inhibit spinal nociceptive information and thereby afferent nociceptive input to the brain. The underlying mechanisms of the various emotional and cognitive modulatory influences may partly overlap, but also have some unique aspects. What becomes clear is that pain is not merely a reflection of the nociceptive input, but should be considered as a complex experience shaped by psychological factors that may be unique for each individual.

Pain is not solely determined by the degree of nociceptive input, but also depends on psychological factors such as emotional state and cognitive appraisal. The International Association for the Study of Pain (IASP) defines pain as 'an unpleasant sensory and emotional experience associated with actual or potential tissue damage, or described in terms of such damage'. According to this definition, pain is unquestionably a bodily sensation, but it is also always unpleasant and therefore an emotional experience [1]. Brain imaging research has demonstrated that sensory, affective and cognitive regions interact to determine the final pain experience. Several decades of research have identified some of the most important emotional and cognitive determinants. This chapter will discuss successively: fear and anxiety, depression, anger and positive affect as the emotional components of pain, and attention, expectancy and appraisal as the cognitive components. Even though for the sake of this overview these aspects are treated separately, as we will see later, some of these components are inherently linked to each other. It should also be kept in mind that although the focus of this chapter is on emotions and cognitions as determinants of the pain experience, there is a reciprocal relationship. Moreover, as implied in the definition, emotions are also an integral part of the pain experience itself.

Emotional Factors

Before proceeding to reviewing emotional modulation of pain, a clarification of terms is in order. *Emotion* is usually conceptualized as a reaction to a particular elicitor and has a clearly defined focus. *Mood* is a more diffuse state without a clear focus. The term *affect* is used to refer to emotions and mood collectively. It should also be noted that emotional factors can refer to short-lasting states (e.g. after anger induction) or to longer-lasting states (e.g. clinical depression). The overview presented below will incorporate studies that have focused on both short- and long-lasting emotional states and moods.

Anxiety and Pain-Related Fear

Research in healthy volunteers using experimental pain stimuli has demonstrated that anxiety, specifically pain-related anxiety, can increase pain sensitivity and exacerbate the pain experience [2–6]. Additionally, in clinical situations anticipatory anxiety has been associated with higher pain reports. For instance, anxiety concerning injections led to higher reported pain after dental injections [7]. Rhudy and Meager [8] distinguished between the effects of fear versus anxiety. Anxiety, defined as the anticipation of potential threats, was proposed to lead to augmentation of the pain experience. Fear, defined as the alarm reaction to a present threat, was proposed to suppress pain experiences and to mobilize the body to allow escape. Their experiment confirmed these predictions: when moderately intense shocks (supposed to elicit fear) were delivered prior to a painful test stimulus, pain was inhibited; however, when threat of intense shocks was induced without actually giving these shocks, sensitivity for the test stimulus was enhanced [8]. The anxiety state would be most characteristic of the emotions experienced in most experimental studies and in clinical situations. Fear and inhibition of pain, on the other hand, may typically be seen in extreme situations such as during combat or assault, where pain reactions might hamper escape and endanger survival. In much of the literature this distinction between anxiety and fear is, however, not made and the term 'pain-related fear' is used to denote both the reaction to current pain as well as anticipatory anxiety for upcoming pain.

Pain modulation by anxiety can be traced back to brain interactions of emotional and sensory areas. Incoming nociceptive signals are processed in many different brain areas such as the somatosensory cortex, anterior cingulate cortex (ACC), insula, amygdala, thalamus and hypothalamus [9, 10]. Projections from the insula and prefrontal cortex, via periaqueductal gray (PAG) to brainstem nuclei, and finally to the spinal cord are involved in descending modulation of pain. This descending modulation can be both inhibitory and excitatory, and involves among other endogenous opioids, noradrenaline and serotonin [9–11].

Anxiety can directly or indirectly influence many different areas involved in coding sensory and affective aspects of pain as well as areas involved in descending modulation. For instance, one study found that conditioned anxiety increased activation of the entorhinal cortex of the hippocampal formation, which subsequently affected activity in the midinsula intensity coding area [12]. Many studies also point to the pivotal role of prefrontal areas, the amygdala and the PAG in conveying the influence of anxiety on pain. Anxiety-prone individuals show impaired prefrontal functioning, which is generally assumed to inhibit pain [9]. In addition, anxious people show a different connectivity pattern between the anterior insular cortex and the PAG, suggestive of less efficient pain modulation [13, 14]. This is not an exhaustive overview of imaging studies that have shown how anxiety might impact on the neural pain network, as a variety of mechanisms have in fact been identified [14].

Anxiety has also been implicated to play a role in chronification of pain. One of the most influential psychological models of chronic pain is the fear-avoidance model [15, 16]. This model has mostly been applied to chronic musculoskeletal pain. According to this model, fear of pain and fear of reinjury may hamper recovery from an acute pain episode because this anticipatory anxiety motivates avoidance behavior and inactivity. Persistent inactivity may be maladaptive and can lead to a negative spiral of deteriorated physical condition, disability and distress, and subsequently to increased pain intensity. Correlational studies have indeed confirmed the predicted relationships between pain-related fear and pain intensity, disability and distress in patients with musculoskeletal pain [17, 18]. More pertinent for its proposed etiological role, prospective studies have shown that pain-related fear in patients seeking care for acute pain may be predictive of pain persistence and long-term disability and sick leave [19–21]. Pain-related fear is also related to the future inception of (back) pain in the general population [22, 23].

A suitable model to study the transition from acute to chronic pain is the postsurgical pain model. Surgery causes tissue injury and massive nociceptive input. Depending on the specific procedure, patients experience acute postsurgical pain for a few hours to several days or even weeks after the operation, which for the majority of patients will subside in time. However, a minority of patients can go on to develop chronic postsurgical pain (CPSP). Prevalence estimates for CPSP have ranged from 5 to 30% for the most common surgical procedures, but can be as high as 50% for amputations and thoracotomy [24, 25]. Because elective surgery is a planned procedure, it allows for the assessment of psychological variables before the intervention takes place. Studies on predictors of CPSP have identified general anxiety as well as surgery-specific anxiety as important predictors of long-term surgical outcome in terms of pain and disability [26, 27].

In conclusion, anxiety seems to aggravate the subjective experience of a nociceptive stimulus, it can trigger maladaptive behaviors in patients with chronic pain and it may predispose people to develop a chronic pain condition. As the next sections will demonstrate, anxiety is certainly not the only emotion affecting the pain experience. However, it is probably the most researched factor in the past decades with generally robust findings.

Depression

A vast number of studies are available demonstrating that negative emotions increase pain perception. Most studies examining the effect of negative emotion induction on experimental pain have not differentiated between the specific emotions that were elicited (e.g. disgust, anger, depression), with a few exceptions. Zelman et al. [28] used the Velten mood induction procedure, where participants read depressive statements, and found lowered tolerance for cold pressor pain in comparison to a neutral mood condition. Berna et al. [29] used the Velten procedure accompanied by sad music to induce a depressed mood and applied tonic heat stimuli in the scanner. Depressed mood induction led to higher perceived pain unpleasantness and increased activity in various pain-related brain areas including prefrontal and anterior cingulate cortices and the hippocampus.

Other studies have examined pain sensitivity and pain processing in patients with major depressive disorders. These studies show an inconsistent relation between clinical depression and pain sensitivity, with hypersensitivity, hyposensitivity or no differences between depressed and nondepressed participants [9]. Brain imaging studies also found contradictory results. For instance Bär et al. [30] found depressive patients have less sensitivity for thermal pain stimuli with concomitant increased activation of the ventrolateral and dorsolateral prefrontal cortices, which have a pain-inhibiting function. On the other hand, Strigo et al. [31] observed decreased activation in pain modulatory areas including prefrontal areas and the PAG during heat pain stimulation.

Clinical observation and epidemiological studies have demonstrated a strong association between chronic pain and depression. As much as 50% of patients with chronic pain suffer from a comorbid depressive disorder [32, 33]. Concurrent depressive symptoms in patients with chronic pain are associated with increased disability and an augmented pain experience [34, 35]. Imaging studies in patients with chronic pain have shown that concomitant depression is especially associated with increased activity in brain regions involved in affective processing both in response to experimental pain and disease-relevant pain [36, 37].

A long-standing debate is whether chronic pain causes depression, or whether depression is a risk factor for developing chronic pain. Both may in fact be true, with one condition constituting a risk factor for the development of the other condition. In addition, the co-occurrence of chronic pain and depression may be due to shared underlying mechanisms, either cognitive behavioral (e.g. repetitive negative thinking, avoidance [38]) or biological (e.g. neuroinflammation [39], genetic [40]). Longitudinal studies provide evidence that having depression might increase the chance of a subsequent pain disorder, which is consistent with the risk factor or shared vulnerability view. The incidence of developing back pain in previously pain-free individuals was found to be 2–3 times higher in individuals who suffered from a previous or current depressive disorder [41, 42]. Also with regard to the development of CPSP, preoperative depression has been identified as a risk factor, although less consistently so compared to anxiety [43, 44].

Anger

Less frequently researched than the above two negative emotions is the influence of anger on pain. Nevertheless, evidence is accumulating that anger can also influence pain perception. Several studies have examined the effects of anger induction on pain sensitivity in pain patients or healthy controls and generally found increased sensitivity for experimental pain stimulation [45, 46]. In addition, anger management style may influence pain sensitivity. Especially individuals who are more prone to express anger in a physical or verbal way (i.e. anger-out style) show greater sensitivity, and this was found for experimental pain, as well as for acute postsurgical pain [47, 48]. It has been speculated that deficient endogenous opioidergic modulation in the rostral limbic system might underlie both increased pain sensitivity and a reduced ability to adequately regulate anger [48]. However, anger management style and state anger can show a complex interaction; when people with a high level of anger cannot express their anger (e.g. in a lab situation it is usually not appropriate to express anger), this may indeed increase sensitivity; however, when anger is expressed, this may in fact normalize or even decrease pain sensitivity [48, 49].

Some indication for a possible causal relation between anger regulation and increased pain comes from daily diary studies with chronic low back pain patients [49, 50]. Behavioral anger expression and anger inhibition were associated with increased pain intensity, more pain interference and decreased functioning at subsequent assessment times. In addition to a direct effect on the pain experience, anger may also have a negative effect on treatment outcome and lead to interpersonal problems [51, 52].

Positive Emotions

Although most research has focused on the role of negative emotions, several studies have specifically examined the effect of positive emotions on pain. It should be noted that positive emotions are not just the opposite of negative emotions, they are correlated but can also vary partly independently [53]. Experimental studies in healthy volunteers have demonstrated that the induction of positive emotions leads to reduced pain sensitivity (for an overview, see Finan and Garland [54]). Again, both supraspinal modulation (i.e. interactions between pain and emotional networks within the brain [14]) and descending pain modulation are involved, as shown by a decrease in a spinal nociceptive reflex after positive emotion induction [55, 56].

Positive emotions may also have a protective role in the transition from acute to chronic pain because they may protect against the development and spreading of pain-related fear. In an elegant experiment, fear for a specific hand movement was experimentally induced by pairing this movement with electrical shock. During the extinction phase, where shocks were no longer given, participants with high trait positive affect showed faster fear reduction than participants with low positive affect [57]. A subsequent study showed that the induction of positive emotions could prevent spreading of fear to new movements and to lower pain expectancy for these movements [58].

In chronic pain patients, positive affect and pain intensity have repeatedly been found to be correlated [59]. However, many of these studies were cross-sectional and did not allow conclusions on the direction of causality. A few studies nevertheless suggested that positive affect may be prospectively related to less intense pain. For instance, a diary study indicated that an increase in positive emotions was related to lower pain reports at subsequent time points in patients with rheumatoid arthritis [60]. Other studies have indicated that positive emotions may buffer the effect of pain on physical functioning and distress. Positive emotions were related to faster functional recovery after hip fractures [61] and to less distress in patients with rheumatoid arthritis at times when pain intensity was high [62]. Induction of positive emotions may reduce clinical pain and distress. A randomized controlled trial studying the effects of an online positive psychology intervention on daily pain demonstrated that people reporting mild-to-moderate pain before the intervention showed significant reductions in pain at the 6-month follow-up [63]. Preliminary data from a study using virtual reality to induce positive emotions suggest that this intervention might decrease pain and distress, and increase activity level in fibromyalgia patients [64].

In sum, positive emotions may have pain-reducing qualities. This effect may at least be partly dependent on their incompatibility with pain-related fear and probably other negative emotions.

Cognitive Factors

In addition to emotional factors, cognitive factors also influence the perception and experience of pain. Cognition is defined as the mental processes related to gaining knowledge and encompasses such functions as attention, memory and reasoning. The cognitive processes most studied in relation to pain are attention, expectancy and appraisal.

Attention

Attention and nociception are intrinsically related and have a bidirectional relationship with each other [65]. Pain can involuntarily capture attention, especially when it is intense, novel or threatening [66]. A shift in attention towards the source of a potential threat enables escape and prevention from further harm. On the other hand, directing attention towards pain and bringing it into focal awareness may increase its intensity. In reverse, directing attention away from pain will exclude it from further processing and lead to less pain [67].

Numerous studies have documented the effect of attention manipulations on perceived intensity of experimental pain, with the general finding that distraction by a competing task diminishes pain and focusing on pain increases it [68, 69]. Brain imaging studies have shown that distraction away from pain is associated with decreased activity in ascending pain pathways, including the thalamus, somatosensory areas, insular cortex and ACC, and increased activity in descending pain modulatory pathways involving the ACC and PAG [70–72]. Animal studies have shown the involvement of dopamine, serotonin and endocannabinoids in attentional modulation of pain [73, 74]. However, involvement of the descending system in attentional control has recently been questioned as some studies suggest that the underlying neural mechanisms of attentional modulation may be partly distinct from those underlying the effects of emotion on pain [72]. When emotion and attention where independently manipulated, emotional influences on pain perception did involve the PAG and ACC, suggestive of descending modulation, but attentional modulation affected the superior posterior parietal cortex and the anterior insula [75]. The latter is consistent with orienting attention toward sensory information and optimizing the cortical presentation of the attended stimulus [74]. Nevertheless, even though attentional and emotional modulations of pain may originate from different cerebral sources and may partly act through different mechanisms, there are also communalities between their actions, e.g. both can modulate activity in the anterior insula [76].

Attentional processes also influence the experience of clinical pain. Heightened attention towards painful signals – also referred to as hypervigilance – have been implied in the transition from acute to chronic pain [16]. Such hypervigilance may especially become apparent in individuals that are highly fearful of pain and for whom avoidance of pain is the main goal [77, 78]. In patients with chronic pain, high pain vigilance was indeed found to be associated with higher pain intensity [79, 80].

Another line of studies has examined the effectiveness of distraction as a coping strategy for managing chronic pain. Although distraction is one of the most frequently used and recommended strategies for managing pain, results regarding its effectiveness remain variable [67]. Several factors may moderate the efficacy of distraction, e.g. motivational relevance of and engagement in the distraction task, pain intensity and habitual hypervigilance [78, 81]. Nevertheless, under proper circumstances, distraction may be a powerful analgesic technique. For instance, the use of virtual reality distraction has proven successful in reducing experimental, procedural and chronic pain [82, 83]. Virtual reality involves multisensory information facilitating full immersion in the distracting environment. Experiencing a high level of presence in the virtual environment, as well as fun in interacting with the environment, determine the efficacy of virtual reality distraction [84].

Expectancy

A plethora of studies have been performed on placebo, and to a lesser extent nocebo, effects on pain, effects that are highly dependent on expectancies of pain relief or pain increases, respectively [85]. Verbal suggestion that pain relief or increase may follow can be sufficient for inducing strong expectations, and subsequently influencing the actual experience of pain [86]. Colloca et al. [87] demonstrated that the verbal suggestion that the intensity of weak electrocutaneous shocks would increase not only led to the judgment of stimuli that were previously just above the pain threshold to be judged as much more painful, but also that tactile stimuli (i.e. stimuli previously judged as nonpainful) could be turned into painful stimuli. Neuroscientific research has indicated that the effects of expectancy on the subjective pain experience are paralleled by changes in afferent nociceptive brain areas (e.g. somatosensory areas and insula) and that this is at least partly mediated by descending pain modulatory circuits (e.g. ACC and PAG) [88–90].

Expectations may also interact with the analgesic efficacy of drugs and with other pain modulatory mechanisms. Induction of positive treatment expectancy was shown to lead to a substantial increase in the analgesic effect of a potent opioid, while induction of a negative treatment expectancy completely abolished the analgesic effect [91]. This effect on pain perception was accompanied by changes in activity in the descending pain modulatory system. Cormier et al. [92] found further evidence that expectancies might work through descending modulation and affect nociceptive processes at the spinal level. They used a conditioned pain modulation procedure, i.e. the phenomenon that a noxious stimulus (the conditioning stimulus) reduces the perceived intensity of another noxious stimulus applied on a different part of the body (the test stimulus). Conditioned pain modulation has been found to depend on descending pathways from the brainstem and to inhibit spinal nociceptive processing [93]. The effect of the conditioning stimulus on pain perception as well as the spinal nociceptive flexion reflex was moderated by expectations. Nociceptive inhibition was enhanced when participants expected the conditioning stimulus to induce analgesia; however, when participants expected more pain after the conditioning stimulus, the prototypical inhibitory effect reversed and increased pain reports and elevated reflexes were found [92].

Despite the fact that placebo and nocebo expectations may act through the same central and spinal mechanism, there may be different underlying biochemical factors [86]. Expectations of pain relief were found to act through activation of the endogenous mu-opioid system [94, 95], whereas expectations of pain increase may be mediated by cholecystokinin [96]. The underlying mechanism of expectation-induced analgesia may also be different from the mechanisms involved in the effects of distraction on pain [97], with expectation relying more on (opioid-dependent) descending pain modulation than distraction [74].

The induction of positive expectancies can also be clinically used, for instance by verbally stressing expected pain relief in patient-healthcare provider interaction. Negative expectancies about the effects of an analgesic treatment – by the patient as well as by the healthcare provider – may on the other hand reduce the efficacy of treatment [98].

Finally, expectancies may influence the recovery from acute pain and injury. Patients' expectations for recovery after traffic-related whiplash injuries have proven to be an important prognostic factor for neck pain intensity and limitations 1 year later [99]. In postoperative pain studies, expected pain and expected recovery were shown to predict actual surgical outcomes. Expectations of high acute postoperative pain were predictive of pain scores during the first postoperative days [100, 101]. Negative expectations about continuing pain, disability and ability to work led to continued pain and disability after lumbar disc surgery for back pain [102, 103]. Also for other operations, such as total knee arthroplasty, negative expectancies were found to be associated with CPSP [104].

Appraisal

More complex cognitive processes, i.e. our interpretation of the meaning of pain, its causes and consequences, also color the experience of pain. Most research has been devoted to one specific (negative) cognitive pain appraisal, catastrophizing. Catastrophizing is defined as an exaggerated negative mental set regarding actual or anticipated pain and encompasses several aspects of negative thinking: exaggeration of the threat value of pain, rumination about pain and feeling helpless in the face of pain [105]. Catastrophizing about pain has close interconnections to some of the concepts discussed above, most notably anxiety, attention and expectancy. According to the fear-avoidance model of pain discussed earlier, catastrophizing is the precursor of pain-related fear and of increased attention towards pain (i.e. pain vigilance) [15]. It also has aspects of expectancy, i.e. expectation that the pain may not go away or becomes worse. Not surprisingly, many of the same associations between catastrophizing and pain have been reported as were found for the constructs discussed above. In healthy individuals, for example, pain catastrophizing is related to increased sensitivity to experimental pain [105, 106], and in chronic patients to higher reported pain intensity, more pain interference and more disability [107]. Also, the pain-modulating influences of catastrophizing seem to be associated with similar brain regions that are implied in attentional and emotional control of pain [108, 109], and with lower activation of descending pain-inhibitory controls [110, 111].

Pain catastrophizing has also been implicated as an important factor in the transition from acute to chronic pain. Catastrophizing predicts the persistence of pain after an acute back pain episode [112] and persistent pain after surgical intervention [27, 113, 114]. A catastrophic thinking style can interact with genetic factors in determining persistent pain. The combination of a specific gene diplotype and high levels of pain catastrophizing were found to increase shoulder pain intensity after exercise-induced muscle injury and shoulder surgery more than either factor on its own [115, 116]. Many more studies in fact found associations between pain catastrophizing and pain outcomes in both healthy volunteers and pain patients; there are several reviews available that summarize this evidence [105, 106, 117].

Another higher-order cognition that has been found to influence pain perception is optimism. Optimism, as well as its reverse pessimism, is related to catastrophizing yet distinct in that it concerns a broader thinking style affecting multiple life domains. It is usually thought of as a trait characteristic, remaining relatively stable across the life span [118]. An optimistic thinking style has been related to less pain sensitivity and to less interference of pain on activities and mood [119]. In laboratory experiments, trait optimism, as well as (temporary) induced optimism, was found to lead to lower perceived pain, and this effect seemed to be mediated by lower levels of pain catastrophizing [120, 121]. Optimism was also indicated as a protective mechanism for the development of chronic postoperative pain after surgery [122].

Concluding Remarks

The above review makes clear that multiple emotional and cognitive factors can impact on the experience of pain. Although the various factors were presented and discussed separately for the sake of this overview, it should be noted that there are profound interconnections between these factors. For instance, both pain catastrophizing and pain-related fear may lead to attention to pain; a general optimistic attitude may result in less fear and less pain catastrophizing, etc. In addition, the underlying pathways may partly overlap. Brain areas processing affective and cognitive information are heavily interconnected and in turn both have connections to sensory areas. Descending pathways play a role in many of the effects presented above. Probably the primary distinction that can be made is between emotion modulation in general (i.e. independent of the specific emotion involved) and attentional modulation because there are indications that for these two modulatory influences the underlying mechanisms may differ. The more complex cognitive factors such as pain catastrophizing and optimism probably rely on multiple underlying emotional (e.g. fear) and more basic cognitive factors (e.g. attention). For treatment and management of (chronic) pain, all of these factors might, however, constitute targets for prevention and intervention, or may be used for determination of risk for a negative pain trajectory. Discussion of treatment strategies for decreasing negative emotions and increasing positive emotions, and/or changing pain cognitions, is beyond the focus of this chapter but can be found elsewhere [123–126]. The principal message is that proper understanding of the experience of pain always involves awareness of emotional and cognitive aspects involved. Pain is a complex experience shaped by multiple factors that are unique for each individual.

References

1 Merskey H, Bogduk N: IASP Task Force on Taxonomy, Classification of Chronic Pain; Part III: Pain Terms, a Current List with Definitions and Notes on Usage. Seattle, IASP Press, 1994.

2 Hirsh AT, George SZ, Bialosky JE, Robinson ME: Fear of pain, pain catastrophizing, and acute pain perception: relative prediction and timing of assessment. J Pain 2008;9:806–912.

3 Horn ME, Alappattu MJ, Gay CW, Bishop M: Fear of severe pain mediates sex differences in pain sensitivity responses to thermal stimuli. Pain Res Treatment 2014;2014:897953.

4 George SZ, Dannecker EA, Robinson ME: Fear of pain, not pain catastrophizing, predicts acute pain intensity, but neither factor predicts tolerance or blood pressure reactivity: an experimental investigation in pain-free individuals. Eur J Pain 2006;10: 457–465.

5 Roelofs J, Peters ML, Deutz J, Spijker C, Vlaeyen JWS: The Fear of Pain Questionnaire (FPQ): further psychometric examination in a non-clinical sample. Pain 2005;116:339–346.

6 Robinson ME, Bialosky JE, Bishop MD, Price DD, George SZ: Supra-threshold scaling, temporal summation, and after-sensation: relationships to each other and anxiety/fear. J Pain Res 2010;3:25–32.

7 van Wijk AJ, Hoogstraten J: Anxiety and pain during dental injections. J Dent 2009;37:700–704.

8 Rhudy JL, Meagher MW: Fear and anxiety: divergent effects on human pain thresholds. Pain 2000;84:65–75.

9 Wiech K, Tracey I: The influence of negative emotions on pain: behavioral effects and neural mechanisms. Neuroimage 2009;47:987–994.

10 Jennings EM, Okine BN, Roche M, Finn DP: Stress-induced hyperalgesia. Prog Neurobiol 2014;121:1–18.

11 Bushnell MC, Čeko M, Low LA: Cognitive and emotional control of pain and its disruption in chronic pain. Nat Rev Neurosci 2103;14:502–511.

12 Ploghaus A, Narain C, Beckmann CF, Clare S, Bantick S, Wise R, Matthews PM, Rawlins NP, Tracey I: Exacerbation of pain by anxiety is associated with activity in a hippocampal network. J Neurosci 2001;21: 9896–9903.

13 Ploner M, Lee MC, Wiech K, Bingel U, Tracey I: Prestimulus functional connectivity determines pain perception in humans. Proc Natl Acad Sci U S A 2010;107:355–360.

14 Roy M, Piche M, Chen J, f, Peretz I, Rainville P: Cerebral and spinal modulation of pain by emotions. Proc Natl Acad Sci U S A 2009;106:20900–20905.

15 Leeuw M, Goossens ME, Linton SJ, Crombez G, Boersma K, Vlaeyen JW: The fear-avoidance model of musculoskeletal pain: current state of scientific evidence. J Behav Med 2007;30:77–94.

16 Vlaeyen JWS, Linton SJ: Fear-avoidance model of chronic musculoskeletal pain: 12 years on. Pain 2012;153:1144–1147.

17 Peters ML, Vlaeyen JWS, Weber WE: The joint contribution of physical pathology, pain-related fear and catastrophizing to chronic pain disability. Pain 2005; 113:45–50.

18 Zale EL, Lange KL, Fields SA, Ditre JW: The relation between pain-related fear and disability: a meta-analysis. J Pain 2013;14:1019–1030.

19 Boersma K, Linton SJ: Screening to identify patients at risk: profiles of psychological risk factors for early intervention. Clin J Pain 2005;21:38–43.

20 Linton SJ: A review of psychological risk factors in back and neck pain. Spine 2000;25:1148–1156.

21 Melloh M, Elfering A, Egli Presland C, Roeder C, Barz T, Rolli Salathé C, Tamcan O, Mueller U, Theis JC: Identification of prognostic factors for chronicity in patients with low back pain: a review of screening instruments. Int Orthop 2009;33:301–313.

22 Linton SJ, Buer N, Vlaeyen JWS, Hellsing AL: Are fear-avoidance beliefs related to the inception of an episode of back pain? A prospective study. Psychol Health 1999;14:1051–1059.

23 Picavet HS, Vlaeyen JWS, Schouten JS: Pain catastrophizing and kinesiophobia: predictors of chronic low back pain. Am J Epidemiol 2002;156:1028–1034.

24 Kehlet H, Jensen TS, Woolf CJ: Persistent postsurgical pain: risk factors and prevention. Lancet 2006; 367:1618–1625.

25 Macrae WA: Chronic post-surgical pain: 10 years on. Br J Anaesth 2008;101:77–86.

26 Peters ML, Sommer M, de Rijke JM, Kessels F, Heineman E, Patijn J, Marcus MA, Vlaeyen JW, van Keef M: Somatic and psychologic predictors of long-term unfavorable outcome after surgical intervention. Ann Surg 2007;245:487–494.

27 Theunissen M, Peters ML, Bruce J, Gramke HF, Marcus MA: Preoperative anxiety and catastrophizing: a systematic review and meta-analysis of the association with chronic postsurgical pain. Clin J Pain 2012;28:819–841.

28 Zelman DC, Howland EW, Nichols SN, Cleeland CS: The effects of induced mood on laboratory pain. Pain 1991;46:105–111.

29 Berna C, Leknes S, Holmes EA, Edwards RA, Goodwin GM, Tracey I: Induction of depressed mood disrupts emotion regulation neurocircuitry and enhances pain unpleasantness. Biol Psychiatry 2010;67: 1083–1090.

30 Bär KJ, Wagner G, Koschke M, Boettger S, Boettger MK, Schlosser R, Sauer H: Increased prefrontal activation during pain perception in major depression. Biol Psychiatry 2007;62:1281–1287.

31 Strigo IA, Simmons AN, Matthews SC, Craig A Paulus MP: Association of major depressive disorder with altered functional brain response during anticipation and processing of heat pain. Arch Gen Psychiatry 2008;65:1275–1284.
32 Bair MJ, Robinson RL, Katon W, Kroenke K: Depression and pain comorbidity. Arch Int Med 2003; 163:2433–2445.
33 Dersh J, Gatchel RJ, Mayer T, Polatin P, Temple OR: Prevalence of psychiatric disorders in patients with chronic disabling occupational spinal disorders. Spine 2006;31:1156–1162.
34 Gaskin, ME, Greene AF, Robinson ME, Geisser ME: Negative affect and the experience of chronic pain. J Psychosom Res 1992;36:707–713.
35 Mangelli L, Gribbin N, Buchi S, Allard S, Sensky T: Psychological well-being in rheumatoid arthritis: relationship to 'disease' variables and affective disturbance. Psychother Psychosom 2002;71:112–116.
36 Giesecke T, Gracely RH, Williams DA, Geisser ME, Petzke FW, Clauw DJ: The relationship between depression, clinical pain, and experimental pain in a chronic pain cohort. Arthritis Rheum 2005;52:1577–1584.
37 Schweinhardt P, Kalk N, Wartolowska K, Chessell I, Wordsworth P, Tracey I: Investigation into the neural correlates of emotional augmentation of clinical pain. Neuroimage 2008;40:759–766.
38 Linton SJ: A transdiagnostic approach to pain and emotion. J Appl Biobehav Res 2013;18:82–103.
39 Walker AK, Kavelaars A, Heijnen CJ, Dantzer R: Neuroinflammation and comorbidity of pain and depression. Pharmacol Rev 2014;66:80–101.
40 Gracely RH, Ceko M, Bushnell MC: Fibromyalgia and depression. Pain Res Treat 2012;2012:486590.
41 Jarvik JG, Hollingworth W, Heagerty PJ, Haynor DR, Boyco EJ, Deyo RA: Three-year incidence of low back pain in an initially asymptomatic cohort. Clinical and imaging risk factors. Spine 2005;30:1541–1548.
42 Larson SL, Clark MR, Eaton WW: Depressive disorder as a long-term antecedent risk factor for incident back pain: a 13-year follow-up study from the Baltimore Epidemiological Catchment Area sample. Psychol Med 2004;34:211–219.
43 Hinrichs-Rocker A, Schulz K, Jarvinen I, Lefering R, Simanski C, Neugebauer EA: Psychosocial predictors and correlates for chronic post-surgical pain (CPSP) – a systematic review. Eur J Pain 2009;13: 719–730.
44 Celestin J, Edwards RR, Jamison RN: Pretreatment psychosocial variables as predictors of outcomes following lumbar surgery and spinal cord stimulation: a systematic review and literature synthesis. Pain Med 2009;10:639–653.
45 Bruehl S, Burns JW, Chung OY, Ward P, Johnson B: Anger and pain sensitivity in chronic low back pain patients and pain-free controls: the role of endogenous opioids. Pain 2002;99:223–233.
46 Middendorp H, Lumley MA, Jacobs JWG, Bijlsma JWJ, Geenen R: The effects of anger and sadness on clinical pain reports and experimentally-induced pain thresholds in women with and without fibromyalgia. Arthritis Care Res (Hoboken) 2010;62: 1370–1376.
47 Bruehl S, Chung OY, Donahue BS, Burns JW: Anger regulation style, postoperative pain, and relationship to the A118G mu opioid receptor gene polymorphism: a preliminary study. J Behav Med 2006;29: 161–169.
48 Bruehl S, Burns JW, Chung OY, Chont M: Pain-related effects of trait anger expression: neural substrates and the role of endogenous opioid mechanisms. Neurosci Biobehav Rev 2009;33:475–491.
49 Burns JW, Gerhart JI, Bruehl S, Peterson KM, Smith DA, Porter LS, Schuster E, Kinner E, Buvanendran A, Fras AM, Keefe FJ: Anger arousal and behavioral anger regulation in everyday life among patients with chronic low back pain: relationships to patient pain and function. Health Psychol 2015;34:547–555.
50 Bruehl S, Liu X, Burns JW, Chont M, Jamison RN: Associations between daily chronic pain intensity, daily anger expression, and trait anger expressiveness: an ecological momentary assessment study. Pain 2012;153:2352–2358.
51 Greenwood KA, Thurston R, Rumble M, Waters SJ, Keefe FJ: Anger and persistent pain: current status and future directions. Pain 2003;103:1–5.
52 Trost Z, Vangronsveld K, Linton SJ, Quartana PJ, Sullivan MJ: Cognitive dimensions of anger in chronic pain. Psychosom Med 2011;73:612–619.
53 Watson D, Tellegen A: Toward a consensual structure of mood. Psychol Bull 1985;98:219–235.
54 Finan PH, Garland EL: The role of positive affect in pain and its treatment. Clin J Pain 2015;31:177–187.
55 Williams AE, Rhudy JL: Affective modulation of eyeblink reactions to noxious sural nerve stimulation: a supraspinal measure of nociceptive reactivity? Int J Psychophysiol 2007;66:255–265.
56 Rhudy JL, Williams AE, McCabe KM, Nguyen MA, Rambo P: Affective modulation of nociception at spinal and supraspinal levels. Psychophysiology 2005;42:579–587.
57 Meulders A, Meulders M, Vlaeyen JWS: Positive affect protects against deficient safety learning during extinction of fear of movement-related pain in healthy individuals scoring relatively high on trait anxiety. J Pain 2014;15:632–644.
58 Geschwind N, Meulders M, Peters ML, Vlaeyen JWS, Meulders A: Can experimentally induced positive affect attenuate generalization of fear of movement related pain? J Pain 2015;16:258–269.

59 Pressman SD, Cohen S: Does positive affect influence health? Psychol Bull 2005;131:925–971.
60 Connelly M, Keefe FJ, Affleck G, Lunley MA, Anderson T, Waters S: Effects of day-to-day affect regulation on the pain experience of patients with rheumatoid arthritis. Pain 2007;131:162–170.
61 Fredman L, Hawkes WG, Black S, Bertrand RM, Magaziner J: Elderly patients with hip fracture with positive affect have better functional recovery over 2 years. J Am Geriatr Soc 2006;54:1074–1081.
62 Zautra AJ, Johnson LM, Davis MC: Positive affect as a source of resilience for women in chronic pain. J Consult Clin Psychol 2005;73:212–220.
63 Hausmann LRM, Parks A, Youk AO, Kwohk CK: Reduction of bodily pain in response to an online positive activities intervention. J Pain 2014;15:560–567.
64 Herrero R, Garcıa-Palacios A, Castilla D, Molinari G, Botella C: Virtual reality for the induction of positive emotions in the treatment of fibromyalgia: a pilot study over acceptability, satisfaction, and the effect of virtual reality on mood. Cyberpsychol Behav Soc Netw 2014;17:379–384.
65 Moriarty O, McGuire BE, Finn DP: The effect of pain on cognitive function: a review of clinical and preclinical research. Prog Neurobiol 2011;93:385–404.
66 Legrain V, Van Damme S, Eccleston C, Davis KD, Seminowicz DA, Crombez G: A neurocognitive model of attention to pain: behavioral and neuroimaging evidence. Pain 2009;144:230–232.
67 Johnson MH: How does distraction work in the management of pain? Curr Pain Headache Rep 2005; 9:90–95.
68 Legrain V, Mancini F, Sambo CF, Torta DM, Ronga I, Valentini E: Cognitive aspects of nociception and pain: bridging neurophysiology with cognitive psychology. Neurophysiol Clin 2012;42:325–336.
69 Van Damme S, Legrain V, Vogt J, Crombez G: Keeping pain in mind: a motivational perspective on attentional processing of pain. Neurosci Biobehav Rev 2010;34:204–213.
70 Tracey I, Ploghaus A, Gati JS, Clare S, Smith S, Menon RS, Matthews PM: Imaging attentional modulation of pain in the periaqueductal gray in humans. J Neurosci 2002;22:2748–2752.
71 Valet M, Sprenger T, Boecker H, Willoch F, Rummeny E, Conrad B, Erhard T, Tolle TR: distraction modulates connectivity of the cingulo-frontal cortex and the midbrain during pain – an fMRI analysis. Pain 2004;109:399–408.
72 Villemure C, Schweinhardt P: Supraspinal pain processing: distinct roles of emotion and attention. Neuroscientist 2010;16:276–284.
73 Ford GK, Moriarty O, McGuire BE, Finn DP: Investigating the effects of distracting stimuli on nociceptive behaviour and associated alterations in brain monoamines in rats. Eur J Pain 2008;12:970–979.
74 Ford GK, Moriarty O, Okine BN, Tully E, Mulcahy A, Harhen B, Finn DP: Involvement of the endocannabinoid system in attentional modulation of nociceptive behaviour in rats. Eur J Pain 2014, DOI: 10.1002/ejp.646.
75 Villemure C, Bushnell MC: Mood influences supraspinal pain processing separately from attention. J Neurosci 2009;29:705–715.
76 Ploner M, Lee MC, Wiech K, Bingel U: Tracey I: Flexible cerebral connectivity patterns subserve contextual modulations of pain. Cereb Cortex 2011;21: 719–726.
77 Crombez G, Van Damme S, Eccleston C: Hypervigilance to pain: an experimental and clinical analysis. Pain 2005;116:4–7.
78 Van Damme S, Legrain V, Vogt J, Crombez G: Keeping pain in mind: a motivational account of attention to pain. Neurosci Biobehav Rev 2010;34:204–213.
79 Crombez G, Eccleston C, Van den Broeck A, Goubert L, Van Houdenhove B: Hypervigilance to pain in fibromyalgia: the mediating role of pain intensity and catastrophic thinking about pain. Clin J Pain 2004;20:98–102.
80 Peters ML, Crombez G: Assessment of attention to pain using handheld computer diaries. Pain Med 2007;8:S110–S120.
81 Van Ryckeghem DM, Crombez G, Van Hulle L, Van Damme S: Attentional bias towards pain-related information diminishes the efficacy of distraction. Pain 2012;153:2345–2351.
82 Botella C, García Palacios A, Baños R, Quero S, Breton-Lopez J: Virtual reality in the treatment of pain. J Cyberther Rehab 2008;1:93–100.
83 Malloy KM, Milling LS: The effectiveness of virtual reality distraction for pain reduction: a systematic review. Clin Psychol Rev 2010;30:1011–1018.
84 Triberti S, Repetto C, Riva G: Psychological factors influencing the effectiveness of virtual reality–based analgesia: a systematic review. Cyberpsychol Behav Soc Networking 2014;17:6.
85 Benedetti F: Mechanisms of placebo and placebo-related effects across diseases and treatments. Annu Rev Pharmacol Toxicol 2008;48:33–60.
86 Benedetti F, Lanotte M, Lopiano L, Colloca L: When words are painful: unraveling the mechanisms of the nocebo effect. Neuroscience 2007;147:260–271.
87 Colloca L, Sigaudo M, Benedetti F: The role of learning in nocebo and placebo effects. Pain 2008;136: 211–218.
88 Keltner JR, Furst A, Fan C, Redfern R, Inglis B, Fields HL: Isolating the modulatory effect of expectation on pain transmission: a functional magnetic resonance imaging study. J Neurosci 2006;26:4437–4443.
89 Koyama T, McHaffie JG, Laurienti PJ, Coghill RC: The subjective experience of pain: where expectations become reality. Proc Natl Acad Sci U S A 2005; 102:12950–12955.

90 Wager TD, Rilling JK, Smith EE, Sokolik A, Casey KL, Davidson RJ, Kosslyn SM, Rose RM, Cohen JD: Placebo-induced changes in FMRI in the anticipation and experience of pain. Science 2004;303: 1162–1167.

91 Bingel U, Wanigasekera V, Wiech K, Mhuircheartaigh RN, Lee MC, Ploner M, Tracey I: The effect of treatment expectation on drug efficacy: imaging the analgesic benefit of the opioid remifentanil. Sci Translat Med 2011;3:70ra14.

92 Cormier S, Piche M, Rainville P: Expectations modulate heterotopic noxious counter-timulation analgesia. J Pain 2013;14:114–125.

93 Bouhassira D, Danziger N: Investigation of brainstem descending pain modulation in animals and humans. Suppl Clin Neurophysiol 2006;58:134–149.

94 Amanzio M, Benedetti F: Neuropharmacological dissection of placebo analgesia: expectation-activated opioid systems versus conditioning-activated specific sub-systems. J Neurosci 1999;19:484–494.

95 Zubieta JK, Bueller JA, Jackson LR, Scott DJ, Xu Y, Koeppe RA, Nichols TE, Stohler C: Placebo effects mediated by endogenous opioid activity on opioid receptors. J Neurosci 2005;25:7754–7762.

96 Benedetti F, Amanzio M, Vighetti S, Asteggiano G: The biochemical and neuroendocrine bases of the hyperalgesic nocebo effect. J Neurosci 2006;26: 12014–12022.

97 Buhle JT, Stevens BL, Friedman JJ, Wager TD: Distraction and placebo: two separate routes to pain control. Psychol Sci 2012;23:246–253.

98 Klinger R, Colloca L, Bingel U, Flor H: Placebo analgesia: clinical applications. Pain 2014;155:1055–1058.

99 Caroll LJ, Holm LW, Ferrari R, Ozegovic D, Cassidy DJ: Recovery in whiplash-associated disorders: do you get what you expect? J Rheumatol 2009;36:1063–1070.

100 Gramke HF, de Rijke JM, van Kleef M, Kessels AG, Peters ML, Sommer M, Marcus MA: Predictive factors of postoperative pain after day-case surgery. Clin J Pain 2009;25:455–460.

101 Sommer M, de Rijke MM, Van Kleef M, Kessels F, Peters ML, Geurts JW, Patijn J, Gramke H, Marcus MA: Predictors of acute postoperative pain after elective surgery. Clin J Pain 2010;26:87–94.

102 den Boer JJ, Oostendorp RA, Beems T, Munneke M, Evers AW: Continued disability and pain after lumbar disc surgery: the role of cognitive-behavioral factors. Pain 2006;123:45–52.

103 Johansson A, Lintom SJ, Rosenblad A, Bergkvist L, Nilssom O: A prospective study of cognitive behavioural factors as predictors of pain, disability and quality of life one year after lumbar disc surgery. Disabil Rehabil 2010;32:521–529.

104 Sullivan S, Tanzer M, Reardon G, Amirault D, Dunbar M, Stanish W: The role of presurgical expectancies in predicting pain and function one year following total knee arthroplasty. Pain 2011;152: 2287–2293.

105 Sullivan MJ, Thorn B, Haythornthwaite J, Keefe F, Martin M, Bradley LA, Lefebvre JC: Theoretical perspectives on the relation between catastrophizing and pain. Clin J Pain 2001;17:52–64.

106 Quartana PJ, Campbell CM, Edwards RR: Pain catastrophizing: a critical review. Expert Rev Neurother 2009;9:745–758.

107 Lamé IE, Peters ML, Vlaeyen JWS, van Kleef M, Patijn J: Quality of life in chronic pain is more associated with beliefs about pain, than with pain intensity. Eur J Pain 2005;9:15–24.

108 Gracely RH, Geisser ME, Giesecke T, Grant MA, Petzke F, Williams DA, Clauw DJ: Pain catastrophizing and neural responses to pain among persons with fibromyalgia. Brain 2004;127:835–843.

109 Seminowicz DA, Davis KD: Cortical responses to pain in healthy individuals depends on pain catastrophizing. Pain 2006;120:297–306.

110 Goodin BR, McGuire L, Allshouse M, Stapleton L, Haythornthwaite JA, Burns N, Mayes LA, Edwards RE: Associations between catastrophizing and endogenous pain-inhibitory processes: sex differences. J Pain 2009;10:180–190.

111 Rhudy JL, Maynard LJ, Russell JL: Does in vivo catastrophizing engage descending modulation of spinal nociception? J Pain 2007;8:325–333.

112 Severeijns R, Vlaeyen JW, van den Hout MA, Picavet HS: Pain catastrophizing and consequences of musculoskeletal pain: a prospective study in the Dutch community. J Pain 2005;6:125–132.

113 Edwards RR, Haythornthwaite JA, Smith MT, Klick B, Katz JN: Catastrophizing and depressive symptoms as prospective predictors of outcomes following total knee replacement. Pain Res Manag 2009;14:307–311.

114 Riddle DL, Wade JB, Jiranek WA, Kong X: Preoperative pain catastrophizing predicts pain outcome after knee arthroplasty. Clin Orthop Relat Res 2010;468:798–806.

115 George SZ, Dover GC, Wallace MR, Sack BK, Herbstman DM, Aydog EM, Fillingim RB: Biopsychosocial influence on exercise-induced delayed onset muscle soreness at the shoulder: pain catastrophizing and catechol-O-methyltransferase (COMT) diplotype predict pain ratings. Clin J Pain 2008;24:793–801.

116 George SZ, Wallace MR, Wright TW, Moser MW, Greenfield WH, Sack BH, Herbstman DM, Fillingim RB: Evidence for a biopsychosocial influence on shoulder pain: Pain catastrophizing and catechol-O-methyltransferase (COMT) diplotype predict clinical pain ratings. Pain 2008;136:53–61.

117 Turner JA, Aaron LA: Pain-related catastrophizing: what is it? Clin J Pain 2001;17:65–71.
118 Carver CS, Scheier MF, Segerstrom SC: Optimism. Clin Psychol Rev 2010;30:879–889.
119 Goodin BR, Bulls HW: Optimism and the experience of pain: benefits of seeing the glass as half full. Curr Pain Headache Rep 2013;17:329–337.
120 Hanssen MM, Peters ML, Vlaeyen JW, Meevissen Y, Vancleef LM: Optimism lowers pain: evidence of the causal status and underlying mechanisms. Pain 2013;154:53–58.
121 Hanssen MM, Vancleef LMG, Vlaeyen JW, Peters ML: More optimism, less pain! The influence of generalized and pain-specific expectations on experienced cold-pressor pain. J Behav Med 2014;37: 47–58.
122 Powell R, Johnston M, Smith WC, King PM, Chambers WA, Krukowski Z, McKee L, Bruce J: Psychological risk factors for chronic post-surgical pain after inguinal hernia repair surgery: a prospective cohort study. Eur J Pain 2012;16:600–610.
123 De Peuter S, de Jong J, Crombez G, Vlaeyen JW: The nature and treatment of pain-related fear in chronic musculoskeletal pain. J Cogn Psychother 2009;23:85–103.
124 Eccleston C, Morley SJ, Williams AD: Psychological approaches to chronic pain management: evidence and challenges. Br J Anaesth 2013;111:59–63.
125 Ehde DM, Dillworth TM, Turner JA: Cognitive-behavioral therapy for individuals With chronic pain efficacy, innovations, and directions for research. Am Psychol 2014;69:153–166.
126 Nicholas MK, Linton SJ, Watson PJ, Main CJ: Early identification and management of psychological risk factors ('yellow flags') in patients with low back pain: a reappraisal. Phys Ther 2011;91:737–753.

Prof. Madelon L. Peters
Department of Clinical Psychological Science, Maastricht University
Universiteitsingel 40
NL–6200 MD Maastricht (The Netherlands)
E-Mail madelon.peters@maastrichtuniversity.nl

Finn DP, Leonard BE (eds): Pain in Psychiatric Disorders. Mod Trends Pharmacopsychiatry.
Basel, Karger, 2015, vol 30, pp 153–165 (DOI: 10.1159/000435939)

Generalized Anxiety Disorder and Pain

Borwin Bandelow

Department of Psychiatry and Psychotherapy, University of Göttingen, Göttingen, Germany

Abstract

In this article, the co-occurrence of anxiety disorders (in particular generalized anxiety disorder) and pain conditions is described, characteristics of chronic pain are explained, and data on the prevalence of co-comorbidity of both conditions are reviewed. Further, hypotheses on the possible psychosocial and neurobiological backgrounds of the high rate of co-occurrence are discussed. This review will also focus on the role of 'unexplained' pain syndromes (e.g. somatic symptom disorder and fibromyalgia) and anxiety. Finally, we address possible treatment strategies for patients with both conditions. There is a need for a rigorous assessment of pain syndromes in generalized anxiety disorder and anxiety in chronic pain conditions in order to prevent subsequent mortality by early treatment of both conditions.

A substantial body of literature supports the association between chronic pain and anxiety. The strong association of medically unexplained pain and generalized anxiety disorder (GAD) has important consequences, such as lower quality of life and higher disability and healthcare resource utilization [1, 2].

Pain is a complex emotion that is determined by sensory as well as social and psychological influences. The presence of a mental disorder may modulate the subjective perception of pain.

Pain may be *acute* or *chronic*. Chronic pain is usually defined as pain that persists for more than 3 months. There are several types of pain: nociceptive (e.g. in response to heat or cold, mechanical force), inflammatory, neuropathic, and functional [3]. Pain can further be divided in two categories: *adaptive* or *maladaptive*. For example, nociceptive pain is a form of adaptive pain in response to a noxious stimulus (e.g. painful heat might keep us away from fire). In contrast, maladaptive pain is often like a false alarm. It is an expression of abnormal sensory processing. Chronic pain is mostly maladaptive. Inflammatory pain may be spontaneous pain or a hypersensitivity to pain in response to tissue damage and inflammation. It can be maladaptive or adaptive (e.g. by preventing us from using a healing limb too much). Neuropathic pain is spontaneous pain and hyperalgesia caused by lesions and dysfunctions in the central and the peripheral nervous system. Causes for neuropathic pain include traumatic lesions (e.g. radiculopathies and carpal tunnel syndrome), metabolic disturbances (e.g. diabetes mellitus, uremia, or hypothyroidism), infections (e.g. postherpetic and HIV), vascular disorders (polyarteriitis nodosa and lupus erythematosus), toxins (alcohol abuse and chemotherapeutics), and vitamin deficiencies (e.g. B_1, B_3, B_6, and B_{12}). Neurological disorders, including lumboischialgia, spinal cord lesions, stroke, multiple sclerosis, or tumors of the central nervous system may lead to central neuropathic pain. Neuropathic pain is often chronic and persists beyond the normal time of healing or for more than 3–6 months [4]. Inflammatory, neuropathic, or noninflammatory/nonneuropathic pain may all be spontaneous or evoked by stimuli. Evoked pain can be either allodynia, i.e. the perception of a normally nonpainful stimulus as painful, or hyperalgesia, i.e. the perception of a normally painful stimulus as more painful than usual.

The term *functional* pain is used when an 'organic' cause of the pain sensation cannot be determined, when the pain is not due to tissue damage, and when it is a maladaptive form of pain. It is assumed that functional pain is caused by abnormal processing in the central nervous system. Examples include somatic symptom disorder with predominant pain, fibromyalgia, tension-type headache, irritable bowel syndrome, noncardiac chest pain in depression, panic disorder, and other anxiety disorders.

Prevalence of General Anxiety Disorder among Treatment-Seeking Chronic Pain Patients

In treatment-seeking patients with chronic pain, the rate of a number of mental disorders is increased. In particular, anxiety disorders are common. Some studies have investigated anxiety levels in clinical samples of pain patients. In 97 men with chronic back pain, lifetime rates of alcohol use disorder, major depression, and anxiety disorders were approximately twice as high as in controls [5]. GAD was prevalent in 13.4% of the patients. In another study, much higher rates of mental disorders were found in chronic than in acute back pain patients. The rate for GAD was 4.0% [6]. In a study with 200 low back pain patients, the most common mental disorders were major depression, substance abuse, and anxiety and personality disorders [7]. The rate of GAD was 2.0%. Symptoms of anxiety were more common in patients with chronic low back pain, especially those with comorbid major depression [8]. In cardiology clinics, many patients with noncardiac chest pain are seen. In one study, elevated anxiety levels were found in patients with noncardiac chest pain in comparison to a healthy reference group [9]. In samples of patients with mixed chronic pain, GAD was prevalent in 4.0–15.2% [10, 11]. In a review by Asmundson and Katz [12], the findings on the co-occurrence of anxiety disorders and pain syndromes are summarized.

In representative samples, higher rates of anxiety symptoms were also found in patients with pain. In an international survey of adults (n = 85,088), mental disorders (mainly mood, anxiety, and substance abuse disorders) were more common among persons with back or neck pain than among persons without these conditions [13]. Data from the US National Comorbidity Survey indicate that the prevalence of anxiety disorders in individuals with chronic arthritic pain, arthritis, migraine, and back pain was elevated relative to the general population [14–16]. It is noteworthy that the associations between the pain conditions and the anxiety disorders were generally larger than those between the pain conditions and depression. Also, in the National Comorbidity Survey with 5,877 respondents, a strong association between anxiety disorders and physical disorders was found [17]. Among specific anxiety disorders, posttraumatic stress disorder, panic attacks, and agoraphobia were more likely to be associated with specific physical disorders than GAD, social phobia, or simple phobia. In a meta-analysis of 41 studies (14 clinical and 27 nonclinical pain), it was found that the construct 'anxiety sensitivity' was strongly associated with fearful appraisals of pain [18]. In the Netherlands Study of Depression and Anxiety, 614 participants with no previous history and no current depression or anxiety at baseline were followed up for 4 years. Patients with pain were found to be at increased risk of developing a first episode of depressive or anxiety disorder. Onset of depression and anxiety was associated with six pain locations (neck, back, head, orofacial area, abdomen, and joints) [19]. In another study from the same working group, a current anxiety disorder and a comorbid depressive and anxiety disorder were more strongly associated with cardiorespiratory pain than gastrointestinal or musculoskeletal pain [20].

Pain in Patients with Generalized Anxiety Disorder

Complaints about pain are more common in GAD patients than in people without an anxiety disorder. In a community survey, painful somatic symptoms were found in 59% of patients who screened positive for GAD. The rate of painful somatic symptoms in this study was even higher (78%) in patients who were comorbid with major depression [21]. In a primary care study, 61% of GAD patients seeking treatment reported significant pain [22]. In a Spanish primary care study, almost all patients (93.5%) who were screened for GAD also reported painful physical symptoms [23]. Neuropathic pain was present in 59.4% of all patients.

In primary care, the diagnosis of GAD is often overseen because the patients do not always reveal that their main problem is an anxiety disorder, but present with somatic symptoms and sleeping problems [24]. In a survey in primary care, all patients of 558 physicians had to complete a questionnaire. Of the patients identified as having GAD, only 13% had anxiety as primary complaint, whereas 29.4% indicated that their primary reason for seeing the physician was a pain condition [25].

The high rate of pain syndromes is not only restricted to GAD, but has also been reported for other anxiety disorders, e.g. panic disorder [26, 27].

Possible Explanations for the Co-Occurrence of Pain and Anxiety

There are different hypotheses on the strong association between anxiety disorders and pain syndromes:

(1) *Co-occurrence of anxiety and pain by chance.* As both pain conditions and anxiety are not rare, they could co-occur in the same patients simply by chance, without any causal relationship among each other.

(2) *Chronic pain causes the anxiety disorder.* After many unsuccessful trials of finding a cause for painful condition or treating it, patients may develop fear of the future [8].

(3) *Anxiety exacerbates pain.* The discomfort associated with anxiety may lower the pain threshold. There is evidence that the fear of pain plays a significant role in the development of chronic pain [28]. Limited evidence has shown that the anxiety disorder preceded the onset of pain in the majority of comorbid cases [10, 12].

(4) *A third factor causes both pain and anxiety.* Some third factor, i.e. a genetic predisposition manifested in neurobiological alterations, shared environmental events, or psychosocial stressors could cause both anxiety and pain.

At present, there is not enough evidence to decide which of these four hypotheses is the most likely, but some evidence points to a common underlying factor. Some neurobiological hypotheses have been developed to explain the co-occurrence of chronic pain and anxiety. One of the main hypotheses on the neurobiology of anxiety disorders is a dysfunction of serotonin and/or norepinephrine neurotransmitter systems. Both anxiety disorders and pain conditions can be treated with drugs that influence serotonergic and/or noradrenergic pathways, such as serotonin or serotonin-noradrenaline reuptake inhibitors. Also, serotonergic and noradrenergic projections descending from the brainstem to the spinal cord may suppress painful input from the body. It was assumed that these may be dysfunctional in depression [29], and the same mechanism may apply for anxiety disorders.

In addition, the hypothalamus-pituitary-adrenal axis has been the target of research. For example, corticotropin-releasing factor-mediated mechanisms in the bed nucleus of the stria terminalis have a pivotal role in stress-induced anxiety and hyperalgesia [30]. Animal data showed that prolonged exposure of the amygdala to elevated corticosteroids facilitates long-term anxiety and pain through activation of glucocorticoid receptors and corticotropin-releasing factor [31].

Also, genetic studies support a common genetic vulnerability for both conditions. Multivariate twin analyses in a population-based study revealed a common pathway model for four functional somatic syndromes (chronic widespread pain, chronic fatigue, irritable bowel syndrome, and recurrent headache) and two psychiatric disorders (major depression and GAD) [32].

Although some evidence supports a common underlying neurobiological/genetic factor as the most likely explanation for the co-occurrence of pain and anxiety disorders, further research is needed to corroborate these findings.

Unexplained Pain

The pain syndromes endorsed by GAD patients are often 'unexplained', i.e. a somatic cause could not be determined. There is a large overlap between somatization, depression, and anxiety [33–35]. Two diagnostic entities, somatic symptom disorder and fibromyalgia, are often found in patients with GAD. In one study examining patients with GAD and panic disorder, a high rate of unexplained medical symptoms was found in both disorders, resulting in overutilization of medical services [36]. There is also a link between chronic fatigue syndrome, GAD, fibromyalgia, and somatization disorder [37]. In a survey including 94,516 participants, it was found that individuals with chronic fatigue syndrome, fibromyalgia, and irritable bowel syndrome had significantly more mood and anxiety disorders than individuals without these functional somatic syndromes [38].

Somatic Symptom Disorder

Patients with this diagnosis have disproportionate and persistent thoughts about the seriousness of their symptoms and a high level of anxiety about their condition. They tend to devote excessive time and energy to their concerns about bodily symptoms. In the DSM-5 [39], the concept of somatoform disorder has been revised. Somatization disorder was changed to somatic symptom disorder. The former diagnosis 'pain disorder' is now called 'somatic symptom disorder with predominant pain'. According to the authors of DSM-5, the previous DSM-IV criteria had overemphasized the centrality of medically unexplained symptoms. Determining that a somatic symptom is medically unexplained may not be reliable, and it may be seen as inappropriate to give an individual a mental disorder diagnosis solely because a medical cause cannot be found. Furthermore, affected patients may regard such a diagnosis that implies their physical symptoms are not 'real' as pejorative and demeaning. The previous diagnosis of somatoform disorders also included patients having 'real' somatic disorders who aggravated their symptoms. In addition, some mental disorders may initially manifest with primarily somatic symptoms (e.g. major depressive disorder,

GAD, or panic disorder). However, unexplained symptoms still remain a key feature in conversion disorder in DSM-5 because it is possible to demonstrate definitively in such disorders that the symptoms are not consistent with medical pathophysiology, according to the authors.

In GAD patients who were followed up for over 40 years in a study by Rubio and Lopez-Ibor [40], it was observed that GAD tended to disappear over the years, but was replaced by somatization disorder.

Fibromyalgia

Fibromyalgia is a symptom complex that has been defined by rheumatologists as a chronic noninflammatory disorder. Symptoms include widespread pain in the muscular and tendon system, increased sensitivity in at least 11 of 19 'tender points' (precisely defined areas in the neck, back, shoulders, hips, and knees), fatigue, unrefreshing sleep, concentration and memory difficulties, pain or cramps in the lower abdomen, depression, and headache [41]. Some of the symptoms such as sleep disturbance, anxiety, fatigue, and concentration difficulties can also occur in GAD. According to the definition of rheumatologists, the condition is not due to a psychiatric disorder, which is in contrast to studies investigating mental disorders in fibromyalgia patients. In one study, psychiatric disorders (mainly anxiety and the mood disorders) were present in 74.8% of the patients with fibromyalgia (n = 115) [42]. In a community sample, the risk of lifetime anxiety disorders was approximately 5-fold higher among women with fibromyalgia, but this was mostly due to the high comorbidity with obsessive-compulsive disorder and posttraumatic stress disorder [43]. The latter disorders are no longer part of the group 'Anxiety Disorders' in DSM-5.

Despite the seemingly preciseness of criteria for fibromyalgia, it is often impossible to distinguish clearly between fibromyalgia and somatic symptom disorder. As both GAD and somatoform disorders/fibromyalgia are associated with considerable medical resource use and economic costs owing to lost work productivity [44, 45], it is not surprising that the economic burden caused by patients suffering from both GAD and somatoform disorders is even higher.

Treatment

Treatment of comorbid pain and anxiety may be complicated by low response rates. For example, in 111 patients who were in a multidisciplinary pain management program for the treatment of chronic musculoskeletal pain, anxiety at baseline predicted significant negative change in health-related quality of life [46].

Possible treatments include psychotherapy, medications, and exercise. The approved indications for the drugs mentioned in this article may vary from country to country. When a drug is used off-label, medicolegal issues have to be taken into account.

Treatment of Generalized Anxiety Disorder

According to treatment guidelines, GAD can be treated with selective serotonin reuptake inhibitors (SSRIs), serotonin norepinephrine reuptake inhibitors (SNRIs), the calcium modulator pregabalin, benzodiazepines, the 5-HT_{1A} agonist buspirone, tricyclic antidepressants, the tricyclic anxiolytic opipramol, and other drugs [47–49]. According to a comprehensive meta-analysis of 234 randomized controlled studies in anxiety disorders, effect sizes were generally larger for drug treatment than for psychotherapy [50].

Treatment of Neuropathic Pain

The European Federation of Neurological Societies Task Force has examined all of the randomized controlled clinical trials that have been performed in the various neuropathic pain conditions since 2005. According to the updated EFNS guidelines [51], the following drugs were recommended as treatments for neuropathic pain:

- Antidepressants (tricyclic antidepressants, the SNRIs duloxetine and venlafaxine)
- Calcium modulators (pregabalin, gabapentin)
- Tramadol and opioids (in various conditions)
- Lidocaine plasters
- Capsaicin patches
- Cannabinoids (for refractory cases of multiple sclerosis and peripheral neuropathic pain)

These recommendations show that psychopharmacological agents play a major role in the treatment of neuropathic pain.

Treatment of Fibromyalgia

Three drugs have been approved by the US Food and Drug Administration on the basis of randomized controlled studies for the treatment of fibromyalgia. These include the calcium modulator pregabalin and the SNRIs duloxetine and milnacipran [52]. A comparison of these three drugs showed equal overall efficacy [53]. However, when looking at single syndromes, duloxetine had superior effects on pain, sleep disturbances, and depressed mood. Pregabalin had a superior effect on pain, sleep disturbances, and fatigue, whereas milnacipran and pregabalin were superior to duloxetine in reducing fatigue. According to a meta-analysis by Okifuji and Hare [52], the evidence for some other medications is preliminary, including tricyclic antidepressants (e.g. amitriptyline), the opioid analgetic tramadol, the dopamine agonist pramipexole, and the cannabinoid nabilone. Other nonapproved treatments include γ-hydroxybutyric acid and sodium oxybate. Lack of efficacy was found for SSRIs, nonsteroidal anti-inflammatory drugs, other opioid analgetics, and benzodiazepines.

Among nonpharmacological treatments, exercise was moderately effective. However, fibromyalgia patients rarely put exercise recommendations into practice. For cognitive behavioral therapy (CBT), effect sizes were small to moderate in a meta-analysis on treatments for fibromyalgia [54]. According to this meta-analysis, a combination of pregabalin or SNRIs with multicomponent therapy, exercise, or CBT was most promising.

Treatment of Somatic Symptom Disorder

Randomized controlled studies on the psychopharmacological or psychological treatment of somatoform and hypochondriacal disorders are scarce, perhaps due to the fact that patients with these conditions often do not accept anything besides 'somatic' treatments. They cannot easily be convinced that they should be treated by a psychiatrist or a psychologist. As a consequence, psychiatric researchers planning to conduct a clinical trial of somatoform disorder have often encountered problems to recruit a sufficient number of probands. Moreover, many affected patients have a tendency to dismiss any treatment as ineffective or even develop new symptoms after initiation of treatment. In addition, pharmaceutical companies may be reluctant to bring a drug for somatoform disorder on the market when it is not guaranteed that a sufficient number of patients will take the drug. Therefore, the number of randomized controlled studies investigating psychological and medications treatments for somatoform disorders is low, although these disorders are common and associated with considerable costs for the health system.

Among psychotherapies, CBT has the best body of evidence. CBT was more effective than no treatment [55, 56] or a wait list [57–61]. However, when CBT was compared to 'treatment as usual' (TAU), only two studies found a difference [62, 63], whereas CBT did not differ from TAU in three trials [64–66]. Psychodynamic therapy was not superior to wait list [61] or TAU [67] in the only available studies. A Cochrane meta-analysis found small effects for all psychological therapies taken together. This analysis did not find effect size differences between CBT and TAU [68].

SSRI antidepressants such as escitalopram [69], fluoxetine [70], and fluvoxamine [71] were effective in placebo-controlled trials, whereas the SNRI venlafaxine was not superior to placebo [72]. The tricyclic anxiolytic opipramol [73] and the neuroleptic drug levosulpiride were effective in placebo-controlled studies [74]; however, these drugs are not available in many countries. St. John's wort showed efficacy in two studies [75, 76]. A Cochrane meta-analysis found evidence showing new-generation antidepressants were effective in reducing the severity of somatoform disorders; however, the overall quality of studies was found to be low [77].

Treatment of Comorbid Cases

Few controlled studies have examined the efficacy of treatments for GAD on pain symptoms (e.g. [78]). However, based on the reviews of evidence-based treatments for neuropathic pain, unexplained pain syndromes such as fibromyalgia and somatoform disorder, and guidelines for the treatment of GAD, treatment suggestions for patients suffering from both conditions can be formulated (table 1).

Table 1. Suggestions for the treatment of patients with GAD and chronic pain conditions

Treatment	Examples	Daily dose, mg
Psychological treatments	CBT	
Medications		
SSRIs	Escitalopram and others	10–20
SNRIs	Duloxetine	60–120
	Venlafaxine	75–225
Calcium modulator	Pregabalin	150–600
Tricyclic antidepressants	Imipramine	75–200
Tricyclic anxiolytic	Opipramol	50–300
Exercise	Aerobic exercise (e.g. running for 5 km three times a week)	

CBT is effective for both anxiety disorders [47] and chronic pain syndromes [79]. Therefore, CBT for patients with GAD and clinically significant pain syndromes should incorporate therapy elements for both conditions. Drugs that have not only shown efficacy in GAD but which are also used to treat neuropathic pain or unexplained pain syndromes may be useful for comorbid patients. Although benzodiazepines are effective in GAD, they are not recommended due to their addiction potential. Also, patients with somatic symptom disorder often tend to abuse benzodiazepines. Analgetics are often not effective in chronic pain conditions. In particular, opioid analgetics should be avoided due to their abuse potential. Patients with somatoform disorders often develop an addiction to these drugs [80].

Aerobic exercise was shown to reduce anxiety in patients with panic disorder [81, 82]. There is some preliminary evidence to support the use of exercise in some categories of musculoskeletal pain [83]. Therefore, exercise may also be considered for patients with both anxiety and chronic pain symptoms.

Conclusions

Many patients are lost in the 'Bermuda Triangle' between anxiety, depression, and pain syndromes. There is a need for a rigorous assessment of pain syndromes in GAD and vice versa in order to prevent subsequent mortality by early treatment of both conditions. Comorbidity of pain and anxiety is associated with both increased disability and increased medical utilization. The high comorbidity can probably be explained by a common underlying factor. Intervention strategies, like psychotherapy, medications, and exercise may reduce suffering and disability associated with co-occurring anxiety and chronic pain.

References

1 Olfson M, Gameroff MJ: Generalized anxiety disorder, somatic pain and health care costs. Gen Hosp Psychiatry 2007;29:310–316.

2 Beesdo K, Hoyer J, Jacobi F, Low NC, Hofler M, Wittchen HU: Association between generalized anxiety levels and pain in a community sample: evidence for diagnostic specificity. J Anxiety Disord 2009;23:684–693.

3 Woolf CJ; American College of Physicians, American Physiological Society: Pain: moving from symptom control toward mechanism-specific pharmacologic management. Ann Intern Med 2004;140:441–451.

4 Woolf CJ, Mannion RJ. Neuropathic pain: aetiology, symptoms, mechanisms, and management. Lancet 1999;353:1959–1964.

5 Atkinson JH, Slater MA, Patterson TL, Grant I, Garfin SR: Prevalence, onset, and risk of psychiatric disorders in men with chronic low back pain: a controlled study. Pain 1991;45:111–121.

6 Kinney RK, Gatchel RJ, Polatin PB, Fogarty WT, Mayer TG: Prevalence of psychopathology in acute and chronic low back pain patients. J Occup Rehabil 1993;3:95–103.

7 Polatin PB, Kinney RK, Gatchel RJ, Lillo E, Mayer TG: Psychiatric illness and chronic low-back pain. The mind and the spine – which goes first? Spine (Phila Pa 1976) 1993;18:66–71.

8 Krishnan KR, France RD, Pelton S, McCann UD, Davidson J, Urban BJ: Chronic pain and depression. II. Symptoms of anxiety in chronic low back pain patients and their relationship to subtypes of depression. Pain 1985;22:289–294.

9 Smeijers L, van de Pas H, Nyklicek I, Notten PJ, Pedersen SS, Kop WJ: The independent association of anxiety with non-cardiac chest pain. Psychol Health 2014;29:253–263.

10 Large RG: DSM-III diagnoses in chronic pain. Confusion or clarity? J Nerv Ment Dis 1986;174:295–303.

11 Fishbain DA, Goldberg M, Meagher BR, Steele R, Rosomoff H: Male and female chronic pain patients categorized by DSM-III psychiatric diagnostic criteria. Pain 1986;26:181–197.

12 Asmundson GJ, Katz J: Understanding the co-occurrence of anxiety disorders and chronic pain: state-of-the-art. Depress Anxiety 2009;26:888–901.

13 Demyttenaere K, Bruffaerts R, Lee S, et al: Mental disorders among persons with chronic back or neck pain: results from the World Mental Health Surveys. Pain 2007;129:332–342.

14 McWilliams LA, Cox BJ, Enns MW: Mood and anxiety disorders associated with chronic pain: an examination in a nationally representative sample. Pain 2003;106:127–133.

15 McWilliams LA, Goodwin RD, Cox BJ: Depression and anxiety associated with three pain conditions: results from a nationally representative sample. Pain 2004;111:77–83.

16 Von Korff M, Crane P, Lane M, et al: Chronic spinal pain and physical-mental comorbidity in the United States: results from the National Comorbidity Survey replication. Pain 2005;113:331–339.

17 Sareen J, Cox BJ, Clara I, Asmundson GJ: The relationship between anxiety disorders and physical disorders in the U.S. National Comorbidity Survey. Depress Anxiety 2005;21:193–202.

18 Ocanez KL, McHugh RK, Otto MW: A meta-analytic review of the association between anxiety sensitivity and pain. Depress Anxiety 2010;27:760–767.

19 Gerrits MM, van Oppen P, van Marwijk HW, Penninx BW, van der Horst HE: Pain and the onset of depressive and anxiety disorders. Pain 2014;155:53–59.

20 de Heer EW, Gerrits MM, Beekman AT, et al: The association of depression and anxiety with pain: a study from NESDA. PLoS One 2014;9:e106907.

21 Romera I, Fernandez-Perez S, Montejo AL, et al: Generalized anxiety disorder, with or without comorbid major depressive disorder, in primary care: prevalence of painful somatic symptoms, functioning and health status. J Affect Disord 2010;127:160–168.

22 Hartford JT, Endicott J, Kornstein SG, et al: Implications of pain in generalized anxiety disorder: efficacy of duloxetine. Prim Care Companion J Clin Psychiatry 2008;10:197–204.

23 Garcia-Campayo J, Caballero F, Perez M, Lopez V: Pain related factors in newly diagnosed generalized anxiety disorder patients. Actas Esp Psiquiatr 2012;40:177–186.

24 Wittchen HU, Hoyer J: Generalized anxiety disorder: nature and course. J Clin Psychiatry 2001;62(suppl 11):15–19; discussion 20–21.

25 Wittchen HU, Kessler RC, Beesdo K, Krause P, Hofler M, Hoyer J: Generalized anxiety and depression in primary care: prevalence, recognition, and management. J Clin Psychiatry 2002;63(suppl 8):24–34.

26 Kuch K, Cox BJ, Woszczyna CB, Swinson RP, Shulman I: Chronic pain in panic disorder. J Behav Ther Exp Psychiatry 1991;22:255–259.

27 Schmidt NB, Santiago HT, Trakowski JH, Kendren JM: Pain in patients with panic disorder: relation to symptoms, cognitive characteristics and treatment outcome. Pain Res Manag 2002;7:134–141.

28 Hasenbring MI, Chehadi O, Titze C, Kreddig N: Fear and anxiety in the transition from acute to chronic pain: there is evidence for endurance besides avoidance. Pain Manag 2014;4:363–374.
29 Stahl SM: The psychopharmacology of painful physical symptoms in depression. J Clin Psychiatry 2002; 63:382–383.
30 Tran L, Schulkin J, Greenwood-Van Meerveld B: Importance of CRF receptor-mediated mechanisms of the bed nucleus of the stria terminalis in the processing of anxiety and pain. Neuropsychopharmacology 2014;39:2633–2645.
31 Tran L, Schulkin J, Ligon CO, Greenwood-Van Meerveld B: Epigenetic modulation of chronic anxiety and pain by histone deacetylation. Mol Psychiatry 2014, Epub ahead of print.
32 Kato K, Sullivan PF, Evengard B, Pedersen NL: A population-based twin study of functional somatic syndromes. Psychol Med 2009;39:497–505.
33 Lowe B, Spitzer RL, Williams JB, Mussell M, Schellberg D, Kroenke K: Depression, anxiety and somatization in primary care: syndrome overlap and functional impairment. Gen Hosp Psychiatry 2008;30: 191–199.
34 Bener A, Al-Kazaz M, Ftouni D, Al-Harthy M, Dafeeah EE: Diagnostic overlap of depressive, anxiety, stress and somatoform disorders in primary care. Asia Pac Psychiatry 2013;5:E29–E38.
35 Lieb R, Meinlschmidt G, Araya R: Epidemiology of the association between somatoform disorders and anxiety and depressive disorders: an update. Psychosom Med 2007;69:860–863.
36 Barbee JG, Todorov AA, Kuczmierczyk AR, et al: Explained and unexplained medical symptoms in generalized anxiety and panic disorder: relationship to the somatoform disorders. Ann Clin Psychiatry 1997;9:149–155.
37 Fischler B, Cluydts R, De Gucht Y, Kaufman L, De Meirleir K: Generalized anxiety disorder in chronic fatigue syndrome. Acta Psychiatr Scand 1997;95: 405–413.
38 Janssens KA, Zijlema WL, Joustra ML, Rosmalen JG: Mood and Anxiety Disorders in Chronic Fatigue Syndrome, Fibromyalgia, and Irritable Bowel Syndrome: Results From the LifeLines Cohort Study. Psychosom Med 2015;77:449–457.
39 APA: Diagnostic and Statistical Manual of Mental Disorders, ed 5 (DSM-5™). Washington, American Psychiatric Association, 2013.
40 Rubio G, Lopez-Ibor JJ: Generalized anxiety disorder: a 40-year follow-up study. Acta Psychiatr Scand 2007;115:372–379.
41 Clauw DJ: Fibromyalgia: a clinical review. JAMA 2014;311:1547–1555.
42 Thieme K, Turk DC, Flor H: Comorbid depression and anxiety in fibromyalgia syndrome: relationship to somatic and psychosocial variables. Psychosom Med 2004;66:837–844.
43 Raphael KG, Janal MN, Nayak S, Schwartz JE, Gallagher RM: Psychiatric comorbidities in a community sample of women with fibromyalgia. Pain 2006; 124:117–125.
44 Hoffman DL, Dukes EM, Wittchen HU: Human and economic burden of generalized anxiety disorder. Depress Anxiety 2008;25:72–90.
45 Sicras-Mainar A, Blanca-Tamayo M, Navarro-Artieda R, Rejas-Gutierrez J: Use of resources and costs profile in patients with fibromyalgia or generalized anxiety disorder in primary care settings (in Spanish). Aten Primaria 2009;41:77–84.
46 Orenius T, Koskela T, Koho P, et al: Anxiety and depression are independent predictors of quality of life of patients with chronic musculoskeletal pain. J Health Psychol 2013;18:167–175.
47 Bandelow B, Zohar J, Hollander E, et al: World Federation of Societies of Biological Psychiatry (WFSBP) guidelines for the pharmacological treatment of anxiety, obsessive-compulsive and post-traumatic stress disorders – first revision. World J Biol Psychiatry 2008;9:248–312.
48 Bandelow B, Lichte T, Rudolf S, Wiltink J, Beutel ME: The German guidelines for the treatment of anxiety disorders. Eur Arch Psychiatry Clin Neurosci 2015;265:363–373.
49 Baldwin DS, Anderson IM, Nutt DJ, et al: Evidence-based pharmacological treatment of anxiety disorders, post-traumatic stress disorder and obsessive-compulsive disorder: a revision of the 2005 guidelines from the British Association for Psychopharmacology. J Psychopharmacol 2014;28:403–439.
50 Bandelow B, Reitt M, Röver C, Michaelis S, Görlich Y, Wedekind D: Efficacy of treatments for anxiety disorders – a meta-analysis. Int Clin Psychopharmacol 2015;30:183–192.
51 Attal N, Cruccu G, Baron R, et al: EFNS guidelines on the pharmacological treatment of neuropathic pain: 2010 revision. Eur J Neurol 2010;17:1113.e88.
52 Okifuji A, Hare BD: Management of fibromyalgia syndrome: review of evidence. Pain Ther 2013;2:87–104.
53 Hauser W, Petzke F, Sommer C: Comparative efficacy and harms of duloxetine, milnacipran, and pregabalin in fibromyalgia syndrome. J Pain 2010;11: 505–521.
54 Nuesch E, Hauser W, Bernardy K, Barth J, Juni P: Comparative efficacy of pharmacological and nonpharmacological interventions in fibromyalgia syndrome: network meta-analysis. Ann Rheum Dis 2013;72:955–962.

55 Lidbeck J: Group therapy for somatization disorders in general practice: effectiveness of a short cognitive-behavioural treatment model. Acta Psychiatr Scand 1997;96:14–24.
56 Bouman TK, Visser S: Cognitive and behavioural treatment of hypochondriasis. Psychother Psychosom 1998;67:214–221.
57 Bleichhardt G, Timmer B, Rief W: Cognitive-behavioural therapy for patients with multiple somatoform symptoms. A randomised controlled trial in tertiary care. J Psychosom Res 2004;56:449–454.
58 Warwick HM, Clark DM, Cobb AM, Salkovskis PM: A controlled trial of cognitive-behavioural treatment of hypochondriasis. Br J Psychiatry 1996;169:189–195.
59 Clark DM, Salkovskis PM, Hackmann A, et al: Two psychological treatments for hypochondriasis. A randomised controlled trial. Br J Psychiatry 1998; 173:218–225.
60 McLeod CC, Budd MA, McClelland DC: Treatment of somatization in primary care. Gen Hosp Psychiatry 1997;19:251–258.
61 Sorensen P, Birket-Smith M, Wattar U, Buemann I, Salkovskis P: A randomized clinical trial of cognitive behavioural therapy versus short-term psychodynamic psychotherapy versus no intervention for patients with hypochondriasis. Psychol Med 2011;41: 431–441.
62 Allen LA, Woolfolk RL, Escobar JI, Gara MA, Hamer RM: Cognitive-behavioral therapy for somatization disorder: a randomized controlled trial. Arch Intern Med 2006;166:1512–1518.
63 Moreno S, Gili M, Magallon R, et al: Effectiveness of group versus individual cognitive-behavioral therapy in patients with abridged somatization disorder: a randomized controlled trial. Psychosom Med 2013; 75:600–608.
64 Rosendal M, Olesen F, Fink P, Toft T, Sokolowski I, Bro F: A randomized controlled trial of brief training in the assessment and treatment of somatization in primary care: effects on patient outcome. Gen Hosp Psychiatry 2007;29:364–373.
65 Kolk AM, Schagen S, Hanewald GJ: Multiple medically unexplained physical symptoms and health care utilization: outcome of psychological intervention and patient-related predictors of change. J Psychosom Res 2004;57:379–389.
66 Speckens AE, van Hemert AM, Spinhoven P, Hawton KE, Bolk JH, Rooijmans HG: Cognitive behavioural therapy for medically unexplained physical symptoms: a randomised controlled trial. BMJ 1995; 311:1328–1332.
67 Sattel H, Lahmann C, Gundel H, et al: Brief psychodynamic interpersonal psychotherapy for patients with multisomatoform disorder: randomised controlled trial. Br J Psychiatry 2012;200:60–67.
68 van Dessel N, den Boeft M, van der Wouden JC, et al: Non-pharmacological interventions for somatoform disorders and medically unexplained physical symptoms (MUPS) in adults. Cochrane Database Syst Rev 2014;11:CD011142.
69 Muller JE, Wentzel I, Koen L, Niehaus DJ, Seedat S, Stein DJ: Escitalopram in the treatment of multisomatoform disorder: a double-blind, placebo-controlled trial. Int Clin Psychopharmacol 2008;23:43–48.
70 Fallon BA, Petkova E, Skritskaya N, et al: A double-masked, placebo-controlled study of fluoxetine for hypochondriasis. J Clin Psychopharmacol 2008;28: 638–645.
71 Turkington D, Grant JB, Ferrier IN, Rao NS, Linsley KR, Young AH: A randomized controlled trial of fluvoxamine in prostatodynia, a male somatoform pain disorder. J Clin Psychiatry 2002;63: 778–781.
72 Kroenke K, Messina N 3rd, Benattia I, Graepel J, Musgnung J: Venlafaxine extended release in the short-term treatment of depressed and anxious primary care patients with multisomatoform disorder. J Clin Psychiatry 2006;67:72–80.
73 Volz HP, Möller HJ: Opipramol in anxiety and somatoform disorders. Results of a controlled study (in German). Fortschr Neurol Psychiatr 1998; 66(suppl 1):S21–S24.
74 Altamura AC, Di Rosa A, Ermentini A, et al: Levosulpiride in somatoform disorders: a double-blind, placebo-controlled cross-over study. Int J Psychiatry Clin Pract 2003;7:155–159.
75 Volz HP, Möller HJ, Reimann I, Stoll KD: Opipramol for the treatment of somatoform disorders results from a placebo-controlled trial. Eur Neuropsychopharmacol 2000;10:211–217.
76 Müller T, Mannel M, Murck H, Rahlfs VW: Treatment of somatoform disorders with St. John's wort: a randomized, double-blind and placebo-controlled trial. Psychosom Med 2004;66:538–547.
77 Kleinstauber M, Witthoft M, Steffanowski A, van Marwijk H, Hiller W, Lambert MJ: Pharmacological interventions for somatoform disorders in adults. Cochrane Database Syst Rev 2014;11: CD010628.
78 Russell JM, Weisberg R, Fava M, Hartford JT, Erickson JS, D'Souza DN: Efficacy of duloxetine in the treatment of generalized anxiety disorder in patients with clinically significant pain symptoms. Depress Anxiety 2008;25:E1–E11.
79 Morley S, Eccleston C, Williams A: Systematic review and meta-analysis of randomized controlled trials of cognitive behaviour therapy and behaviour therapy for chronic pain in adults, excluding headache. Pain 1999;80:1–13.

80 Wambach S, Rohr P, Häuser W: Abuse of opioid therapy in somatoform pain disorder. A contribution from a psychosomatic/pain therapist point of view to the discussion of the indication of opioids in nonmalignant pain based on 8 cases (in German). Schmerz 2001;15:254–264.
81 Broocks A, Bandelow B, Pekrun G, et al: Comparison of aerobic exercise, clomipramine, and placebo in the treatment of panic disorder. Am J Psychiatry 1998;155:603–609.
82 Wedekind D, Broocks A, Weiss N, Engel K, Neubert K, Bandelow B: A randomized, controlled trial of aerobic exercise in combination with paroxetine in the treatment of panic disorder. World J Biol Psychiatry 2010;11:904–913.
83 Wright A, Sluka KA: Nonpharmacological treatments for musculoskeletal pain. Clin J Pain 2001;17: 33–46.

Prof. Dr. Borwin Bandelow
Department of Psychiatry and Psychotherapy, University of Göttingen
Von-Siebold-Strasse 5
DE–37075 Göttingen (Germany)
E-Mail bbandel@gwdg.de

Finn DP, Leonard BE (eds): Pain in Psychiatric Disorders. Mod Trends Pharmacopsychiatry.
Basel, Karger, 2015, vol 30, pp 166–175 (DOI: 10.1159/000435940)

Pain in Borderline Personality Disorder

Christian Schmahl[a] · Ulf Baumgärtner[b]

[a]Department of Psychosomatic Medicine and Psychotherapy, Central Institute of Mental Health, Medical Faculty Mannheim/Heidelberg University, and [b]Department of Neurophysiology, Center for Biomedicine and Medical Technology Mannheim (CBTM), Medical Faculty Mannheim/Heidelberg University, Mannheim, Germany

Abstract

Pain processing in patients with borderline personality disorder (BPD) is abnormal primarily with respect to pain thresholds which are typically elevated or perception of phasic nociceptive stimuli which is reduced. In spite of this common finding, nonsuicidal self-injury (NSSI), often expressed as cutting, is a hallmark sign of the disease and serves to release aversive inner tension. The question thus arises, how does a painful stimulus release inner tension when these patients feel less pain than healthy people? However, intensity discrimination is normal in these patients. Imaging data have provided evidence that inhibitory top-down modulation is increased in BPD patients, and that processing of the affective-emotional pain component is altered. Recent studies have focused on the role of pain, tissue injury and seeing blood in the context of NSSI. Preliminary findings suggest a significant role of pain irrespective of concomitant tissue injury, and of seeing blood expressed as a stronger immediate stress release. Taken together, BPD patients exhibit altered pain processing that can be assigned to altered processing of nociceptive stimuli in prefrontal and limbic brain areas, which may help to mechanistically explain the clinical behavior.

Borderline personality disorder (BPD) is a severe psychiatric disorder with a prevalence of 2–3%. According to current conceptualizations, psychopathology of BPD is related to the three core domains: (1) disturbed emotional processing and emotion dysregulation, (2) behavioral dysregulation and impulsivity, and (3) interpersonal disturbances [1, 2]. Previous research in individuals with BPD points to an interaction between dysfunctional affective and cognitive appraisal processes, maladaptive behavior patterns, and neurobiological alterations such as an imbalanced network of frontolimbic brain regions [2–4]. A very common dysfunctional behavior, which is closely linked to emotion dysregulation in BPD, is nonsuicidal self-injury (NSSI) [5].

Interestingly, many BPD patients report analgesic phenomena during self-injury [6]. Since pain appears to have an important emotion-regulating effect in BPD [7], it is crucial to explore the role of pain as well as the interaction between pain and emotions at the neurobiological level.

A large body of experimental research investigating pain perception in individuals with BPD point to reduced sensitivity in BPD patients [8–15]. Preliminary evidence from recent studies indicates specificity of reduced pain sensitivity, as no differences in proprioception and exteroceptive sensitivity have been found [16]. Pain processing can be separated into sensory-discriminative, affective-motivational, and cognitive-evaluative components with different contributions of the individual components to disturbance of pain processing in psychiatric disorders [17].

According to Melzack and Casey [18], the nociceptive system can be separated into two functional-anatomical parts. The lateral system, comprising the lateral part of the thalamus (specifically the VPM and VPL nuclei) as well as the primary and secondary somatosensory cortices (S1, S2), is thought to be primarily involved in processing the sensory aspects of pain, e.g. intensity coding and spatial discrimination. The medial system, comprising the medial thalamic nuclei, the anterior and mid-cingulate cortex, as well as other limbic structures like the prefrontal cortex and amygdala, is associated with processing of the affective pain component, where the actual pain is compared with previous events (involving memory function structurally based in hippocampal, parahippocampal, and amygdala subregions). Depending on the results of this comparison or evaluation, a response is primed in limbic and cingulate motor areas, yielding the adequate reaction under the given circumstances like a 'fight or flight' reaction. The insular cortex plays an important role in the processing of nociceptive stimuli and ongoing pain although the insula cannot be assigned clearly to the lateral or medial system since it processes both aspects of the pain experience. The posterior part (dorsal posterior insula) contains a somatotopic representation for nociceptive stimuli in the anterior-posterior direction (hand anterior, foot posterior) for superficial and deep (muscle) pain [19–22]. Cortical stimulation in epilepsy patients only in the posterior insula and medial parts of the operculum leads to the perception of pain in different areas of the body [23, 24]. Hence, the posterior insula could be assigned to the sensory/lateral system, with properties characterizing a 'primary' nociceptive area. The anterior and mid-insula are involved in autonomic regulation, receive strong limbic input, and show increased activity in neuropathic pain patients and tonic rather than phasic pain. Thus, the anterior insula could be assigned to the affective/medial system. According to Craig [25], the posterior part is involved in exteroceptive function and the anterior insula in interoceptive function, more related to one's 'self' and can be seen as a homeostatic control center of the body. Impairment of insular function may result both in decreased pain sensitivity or increased pain sensitivity as part of a larger network. Lesions in the insula are typically related to decreased thermosensory function with higher thresholds [26, 27]. On the other hand, an irritable cortical focus in the insular cortex may result in intractable epileptic pain attacks [28]. Following stroke, the insular cortex can be involved in the generation of central poststroke pain [29]. Recent neuroimaging studies in BPD have demonstrated structural and functional abnormalities in a frontolimbic network including the cingulate cortex, insula, and amygdala with their well-known connections [30] as a putative correlate of disturbed processing of emotional aspects of pain [31] (see below).

To investigate the contribution of sensory and affective aspects of pain processing in BPD, we applied laser-evoked pain and a spatial discrimination task to test both pain perception and thresholds. In this study, subjective ratings and electroencephalography were measured [11]. In accordance with previous findings, we were able to show that subjective pain ratings were lower in BPD patients and that pain thresholds were altered. Interestingly, no differences were observed in discrimination task performance of laser-evoked brain potentials. Accordingly, the analgesic state in BPD cannot be explained by an impairment of the sensory-discriminative component of pain [11]. A modulation of the affective pain component in BPD patients by the well-known COMT val158met polymorphism has been demonstrated [32], but this finding needs replication.

Aiming to explore neural correlates of reduced pain sensitivity in BPD, several studies have used fMRI to assess brain activation in response to painful stimulation. In one study [15], patients with BPD and NSSI such as cutting, burning, or head-banging underwent an fMRI scan while heat stimuli were applied to their hands. Although BPD patients showed increased pain thresholds and smaller overall activation in response to *standardized* temperature stimuli (compared to healthy controls), the overall activation was similar in response to *individually adjusted* heat stimuli. However, BPD patients showed increased dorsolateral prefrontal cortex (DLPFC) activation along with decreased activations of the posterior parietal cortex. Moreover, painful heat stimulation evoked neural deactivation in the amygdala and the perigenual anterior cingulate cortex (ACC) in participants with BPD [15]. The observed interaction between increased pain-induced response in the DLPFC coupled with deactivations in the ACC and amygdala could be interpreted as an antinociceptive mechanism in BPD [15]. While sensory discrimination processes remain intact, this mechanism may modulate pain circuits primarily by an increased top-down regulation of emotional components of pain or an altered affective appraisal of pain [15].

Based on these initial findings of altered pain processing in BPD, Kraus et al. [33] investigated whether the observed antinociceptive mechanisms distinguished BPD patients with co-occurrent posttraumatic stress disorder (PTSD) from those without PTSD. While no differences in reported pain sensitivity were found between the two patient groups, fMRI results revealed that deactivation in the right amygdala was less pronounced in BPD patients without comorbid PTSD compared to the BPD group with PTSD. Moreover, this deactivation was neither associated with BPD symptom severity nor state dissociation. Based on these data, the authors speculated that the degree of amygdala deactivation during pain processing is modulated by comorbid PTSD in BPD patients.

Examining the neural mechanisms underlying the role of self-inflicted pain as a means of affect regulation in BPD more directly, Niedtfeld and colleagues [34] conducted an fMRI study using picture stimuli to elicit negative affect and thermal stimuli to induce heat pain. Although the authors found an attenuation of activation in limbic areas (amygdala, insula) in response to sensory stimulation, it was neither specific to patients with BPD nor to painful stimulation. The authors argue that possible mechanisms causing amygdala deactivation over time could include an attentional shift caused by sensory stimuli per se [35], the automatic use of cognitive regulation strategies or (re-) appraisal [36], or habituation processes [37]. In order to identify brain mechanisms potentially underlying the limbic deactivation observed in their initial data analysis [34], the authors re-analyzed their findings of pain-mediated emotion regulation in BPD by measuring functional connectivity using psychophysiological interaction analyses. Guided by the intention to reveal potential neural mechanisms responsible for the role of pain in affect regulation in BPD, their analyses focused on connectivity between those regions that had previously been identified to be involved in emotion processing, namely the amygdala, insula, and ACC [38]. Results of their analyses indicated that there is a modulating effect of pain on the processing of negative pictures in BPD patients. More precisely, they observed that painful sensory stimuli, as opposed to warmth perception, resulted in enhanced negative coupling between (para-) limbic and prefrontal structures in BPD, thus indicating inhibition of limbic arousal [38]. Moreover, the authors stated that in healthy participants this pattern was only noticeable when negative pictures were paired with warm stimuli and interpreted this finding as the result of automatic emotion regulation processes in response to negative states of arousal [36]. It is important to note that the statistically significant clusters in the prefrontal cortex, related to enhanced inhibitory coupling in BPD under pain, were located in close spatial proximity in the right medial frontal gyrus [Brodman area (BA) 8 and BA9], i.e. in regions previously associated with emotion regulation processes. For instance, a study by McRae et al. [39] found that BA8 was activated in the course of attentional shift processes, while activation in the DLPFC (BA9) was observed when participants were asked to reappraise presented stimuli [40, 41]. It is also interesting to pay attention to the observation of Niedtfeld et al. [38] that the DLPFC showed enhanced coupling to the posterior insula. As already mentioned above, the insular cortex is known to play a critical role in pain processing [42] as well as in the affective appraisal of pain perception [43]. Hence, one could speculate that the connectivity between the DLPFC and posterior insula in BPD may imply an altered, possibly positive, appraisal of pain, whereas connectivity between BA8 and the amygdala and ACC may reflect a neuronal correlate of attentional shift.

Additional evidence for an altered appraisal mechanism of pain in patients with BPD stems from a recent fMRI study by Kluetsch et al. [44]. Analyses of functional connectivity in brain activation during pain processing in BPD showed that patients with BPD exhibited less connectivity between the posterior cingulate cortex and DLPFC when exposed to painful heat rather than neutral temperature stimulation. Moreover, Kluetsch et al. [44] reported that participants with BPD showed reduced integration of the left retrosplenial cortex, the right inferior temporal gyrus, and the left superior frontal gyrus in the default mode network. Thus, the authors discuss their findings as a possible indicator for a cognitive and affective appraisal of pain in patients with BPD that is less self-relevant and aversive. In sum, the overall reported results on pain processing in BPD suggest that NSSI may be interpreted as an attempt to compensate for a deficient emotion regulation mechanism in BPD patients. More specifically, the reported findings implicate that the soothing effect of pain in BPD seems to be mediated by different emotion regulation processes (attentional shift and altered appraisal of pain).

A very common dysfunctional behavior closely linked to emotion dysregulation in BPD is NSSI [5]. In a dysfunctional attempt to regulate aversive emotional states, a high percentage of BPD patients (69–90%) [45] show NSSI, mainly in the form of skin cutting or burning [7, 46–48]. The most prominent motives for NSSI in these patients are a reduction of aversive inner tension and dissociation [46, 47]. Nock and Prinstein [48] in their model of NSSI distinguish between automatic (subjective) and social reinforcement of NSSI, with subdifferentiation of positive and negative reinforcement. Reduction of unpleasant feelings or tension by NSSI constitutes an example of automatic negative reinforcement. In an extension of this model, Chapman et al. [46] suggest that NSSI primarily serves to avoid unpleasant experiences and feelings, leading to a temporary relief.

From a pathophysiological point of view, an injury would cause both stress and pain in healthy subjects, with pain on top being a trigger for even more stress. Stress by itself reduces pain perception and pain-related behavior in animals where it is termed 'stress-induced analgesia'. This interaction is altered in most BPD patients (fig. 1), where NSSI is used as vehicle to *reduce* aversive inner tension with the 'side effect' of pain, bringing the patients back to lower tension levels. It is unclear what the main underlying mechanism is. It could be either due to a direct paradoxical feedback mechanism between (self-) injury and stress with stress decrease following injury through autonomic-limbic pathways, or could be due to the pain experience associated with the injury with a paradoxical feedback there (pain leads to a *decrease* in stress) in the nociceptive and limbic-behavioral networks, or both. In particular, those patients with BPD (and also other psychiatric disorders) who frequently injure themselves exhibit elevated pain thresholds. This correlation may be due to adaptive processes in central nervous system areas processing nociception that may be termed as central habituation. This tonic influence is kept as a trait by ongoing self-injurious behavior and is less likely the result of phasic pre-

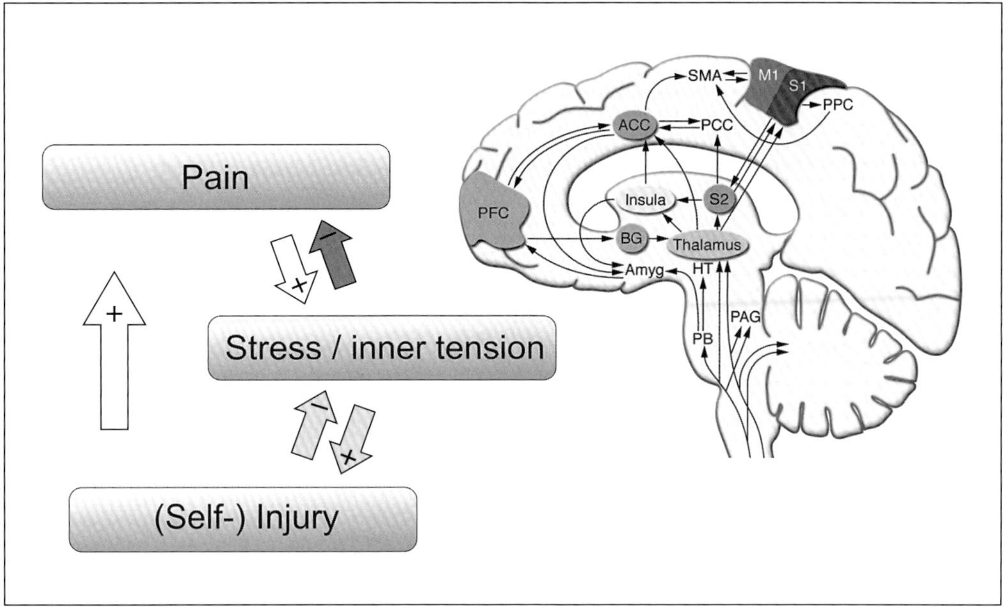

Fig. 1. Concept sketch for normal interaction of stress/inner tension, injury, and pain in healthy subjects (white arrows) and BPD patients (light grey arrows), or both groups (dark grey arrow). Normally, injury directly leads to stress, as well as to pain, and through pain again to stress. BPD patients, however, use self-injury as vehicle to reduce aversive stress. The beneficial effect of self-injury on stress levels in patients may be due to direct feedback between stress and injury, or through pain, which is associated with the injury, or both mechanisms may play an important role. Amyg = Amygdala; BG = basal ganglia; PFC = prefrontal cortex; M1 = primary motor cortex; S1 and S2 = primary and secondary somatosensory cortex; PCC = posterior cingulate cortex; SMA = supplementary motor area; PPC = posterior parietal cortex; HT = hypothalamus; PB = parabrachial plexus; PAG = periaqueductal grey.

frontal cortex hyperactivity with, as a second step, top-down inhibitory modulation of insular activity and blocking of incoming nociceptive signals from the spinal cord dorsal horn.

An important aspect of NSSI is tissue injury which is part of the most frequently used methods such as cutting or burning. We therefore used the model of incision-induced pain, which takes into account tissue injury and thus provides a more valid model for NSSI. In a pilot study of 14 patients with BPD and 18 healthy controls [49], stress induction was followed by an incision into the forearm with a scalpel conducted by an investigator. For stress induction, we used the Montreal Imaging Stress Task (MIST), which combines mental arithmetic with time pressure and negative social feedback [50]. In controls, the incision resulted in a further increase of aversive tension, whereas tension and heart rate decreased in the BPD group after incision. These findings were recently replicated in an fMRI study [51]. In this investigation, 21 female patients with BPD and 17 healthy subjects underwent stress induction with the

MIST, again followed by either an incision into the forearm or sham treatment, where the arm was nonpainfully touched with the blunt end of the scalpel. Participants underwent resting-state fMRI while aversive tension, heart rate, and heart rate variability were assessed. We found a significant influence of incision on subjective and objective stress levels with a stronger decrease of aversive tension in BPD patients following incision as compared to sham. As a neural correlate of the differential processing of incision pain in healthy volunteers and BPD patients, we found that amygdala activity decreased more strongly after incision than after sham treatment in patients, while the opposite pattern was found in healthy controls. Furthermore, functional connectivity of the amygdala with superior frontal gyrus normalized after incision in BPD patients. Our findings of decreased stress levels and amygdala activity after incision support the assumption of an influence of NSSI on emotion regulation in BPD and may help in the understanding of why these patients use self-inflicted pain to reduce inner tension.

Preliminary results of current studies indicate additive roles of pain with and without tissue injury, and of visual cues like seeing blood in the context of injury. In order to dissect the effect of tissue injury associated with pain on stress release and the effect of just pain without tissue injury, we recently established a surrogate model for incision pain that causes the same amount of pain as incision into the skin (model for postoperative pain) [52], but without tissue damage. The custom-made stimulator was a hand-held mechanical stimulator with a tube containing a freely moving weight that was attached to a tip with a 'dull blade' (contact area: 0.1 × 4.0 mm). In healthy subjects, the time course of perceived pain during a 7-second application to the forearm was almost identical to the initial time course of pain elicited by incision pain [53]. Using this novel surrogate model, we compared the impact of tissue injury using a scalpel [52] with the noninvasive surrogate model ('blade stimulus') on stress release in an experimental study in BPD patients and controls. After stress induction with the MIST, both real incision and the blade stimulus led to an immediate decrease of subjective stress, indicating pain as a common mechanistic principle under these circumstances. The impact of seeing blood at the skin site of (pseudo-) tissue damage with the same surrogate 'blade' stimulus seems to have a relevant impact on stress release in BPD patients only [54]. Further studies on the mechanisms behind NSSI are underway to substantiate these data.

Taken together, there is clear evidence of reduced pain sensitivity in patients with BPD. Still, it appears paradoxical why these patients use self-inflicted pain to regulate their emotions in spite of this reduction in pain sensitivity. It appears that pain stimuli of high intensity, e.g. during cutting behavior, serve as potent modulators of stress in this population. Recent research has clearly demonstrated differences between BPD patients and healthy controls in the effects of painful stimuli on experimentally induced stress on behavioral as well as neural levels. Prefrontal-limbic connections appear to play a decisive role in the interaction between pain, stress, and NSSI.

Preliminary studies trying to disentangle NSSI mechanisms point to a significant role of pain irrespective of concomitant tissue injury and of seeing blood for stronger immediate stress release. Further research is needed to elucidate the interaction between pain and NSSI, and to develop therapeutic tools to help BPD patients overcome this devastating behavior.

References

1 Lieb K, Zanarini MC, Schmahl C, Linehan MM, Bohus M: Borderline personality disorder. Lancet 2004; 364:453–461.

2 Leichsenring F, Leibing E, Kruse J, New AS, Leweke F: Borderline personality disorder. Lancet 2011;377: 74–84.

3 Lis E, Greenfield B, Henry M, Guile JM, Dougherty G: Neuroimaging and genetics of borderline personality disorder: a review. J Psychiatry Neurosci 2007; 32:162–173.

4 O'Neill A, Frodl T: Brain structure and function in borderline personality disorder. Brain Struct Funct 2012;217:767–782.

5 Welch SS, Linehan MM, Sylvers P, Chittams J, Rizvi SL: Emotional responses to self-injury imagery among adults with borderline personality disorder. J Consult Clin Psychol 2008;76:45–51.

6 Shearer SL: Phenomenology of self-injury among inpatient women with borderline personality disorder. J Nerv Ment Dis 1994;182:524–526.

7 Klonsky ED: The functions of deliberate self-injury: a review of the evidence. Clin Psychol Rev 2007;27: 226–239.

8 Russ MJ, Roth SD, Lerman A, Kakuma T, Harrison K, Shindledecker RD, et al: Pain perception in self-injurious patients with borderline personality disorder. Biol Psychiatry 1992;32:501–511.

9 McCown W, Galina H, Johnson J, de Simone PA, Posa J: Borderline personality disorder and laboratory-induced cold pressor pain: evidence of stress-induced analgesia. J Psychopathol Behav Assess 1993; 15:87–95.

10 Bohus M, Limberger M, Ebner U, Glocker FX, Schwarz B, Wernz M, et al: Pain perception during self-reported distress and calmness in patients with borderline personality disorder and self-mutilating behavior. Psychiatry Res 2000;95:251–260.

11 Schmahl C, Greffrath W, Baumgartner U, Schlereth T, Magerl W, Philipsen A, et al: Differential nociceptive deficits in patients with borderline personality disorder and self-injurious behavior: laser-evoked potentials, spatial discrimination of noxious stimuli, and pain ratings. Pain 2004;110:470–479.

12 Ludäscher P, Bohus M, Lieb K, Philipsen A, Schmahl C: Elevated pain thresholds correlate with dissociation and aversive arousal in patients with borderline personality disorder. Psychiatry Res 2007;149:291–296.

13 Schmahl C, Meinzer M, Zeuch A, Fichter M, Cebulla M, Kleindienst N, et al: Pain sensitivity is reduced in borderline personality disorder, but not in posttraumatic stress disorder and bulimia nervosa. World J Biol Psychiatry 2010;11:364–371.

14 Cardenas-Morales L, Fladung AK, Kammer T, Schmahl C, Plener PL, Connemann BJ, et al: Exploring the affective component of pain perception during aversive stimulation in borderline personality disorder. Psychiatry Res 2011;186:458–460.

15 Schmahl C, Bohus M, Esposito F, Treede RD, Di Salle F, Greffrath W, et al: Neural correlates of antinociception in borderline personality disorder. Arch Gen Psychiatry 2006;63:659–667.

16 Pavony MT, Lenzenweger MF: Somatosensory processing and borderline personality disorder: pain perception and a signal detection analysis of proprioception and exteroceptive sensitivity. Personal Disord 2014;5:164–171.

17 Klossika I, Flor H, Kamping S, Bleichhardt G, Trautmann N, Treede RD, et al: Emotional modulation of pain: a clinical perspective. Pain 2006;124:264–268.

18 Melzack R, Casey KL: Sensory, motivational, and central control determinants of pain; in Kenshalo DR (ed): The Skin Senses. Springfield, Charles C. Thomas, 1968, pp 423–439.

19 Baumgärtner U, Iannetti GD, Zambreanu L, Stoeter P, Treede R-D, Tracey I: Multiple somatotopic representations of heat and mechanical pain in the operculo-insular cortex: a high-resolution fMRI study. J Neurophysiol 2010;104:2863–2872.

20 Brooks J, Zambreanu L, Godinez A, Tracey I: Somatotopic organisation of the human insula to painful heat studied with high resolution functional imaging. Neuroimage 2005;27:201–209.

21 Henderson L, Gandevia S, Macefield V: Somatotopic organization of the processing of muscle and cutaneous pain in the left and right insula cortex: a single-trial fMRI study. Pain 2007;128:20–30.

22 Henderson LA, Rubin TK, Macefield VG: Within-limb somatotopic representation of acute muscle pain in the human contralateral dorsal posterior insula. Hum Brain Mapp 2011;32:1592–1601.
23 Mazzola L, Isnard J, Peyron R, Guenot M, Mauguiere F: Somatotopic organization of pain responses to direct electrical stimulation of the human insular cortex. Pain 2009;146:99–104.
24 Mazzola L, Isnard J, Peyron R, Mauguière F: Stimulation of the human cortex and the experience of pain: Wilder Penfield's observations revisited. Brain 2012;135:631–640.
25 Craig AB: How do you feel – now? The anterior insula and human awareness. Nat Rev Neurosci 2009; 10:59–70.
26 Greenspan JD, Lee RR, Lenz FA: Pain sensitivity alterations as a function of lesion location in the parasylvian cortex. Pain 1999;81:273–282.
27 Starr CJ, Sawaki L, Wittenberg GF, Burdette JH, Oshiro Y, Quevedo AS, et al: Roles of the insular cortex in the modulation of pain: insights from brain lesions. J Neurosci 2009;29:2684–2694.
28 Isnard J, Magnin M, Jung J, Mauguière F, Garcia-Larrea L: Does the insula tell our brain that we are in pain? Pain 2011;152:946–951.
29 Garcia-Larrea L, Perchet C, Creac'h C, Convers P, Peyron R, Laurent B, et al: Operculo-insular pain (parasylvian pain): a distinct central pain syndrome. Brain 2010;133:2528–2539.
30 Shi CJ, Cassell M: Cortical, thalamic, and amygdaloid connections of the anterior and posterior insular cortices. J Comp Neurol 1998;399:440–468.
31 Krause-Utz A, Winter D, Niedtfeld I, Schmahl C: The latest neuroimaging findings in borderline personality disorder. Curr Psychiatry Rep 2014;16:1–13.
32 Schmahl C, Ludascher P, Greffrath W, Kraus A, Valerius G, Schulze TG, et al: COMT val158met polymorphism and neural pain processing. PLoS One 2012;7:e23658.
33 Kraus A, Esposito F, Seifritz E, Di Salle F, Ruf M, Valerius G, et al: Amygdala deactivation as a neural correlate of pain processing in patients with borderline personality disorder and co-occurrent posttraumatic stress disorder. Biol Psychiatry 2009;65:819–822.
34 Niedtfeld I, Schulze L, Kirsch P, Herpertz SC, Bohus M, Schmahl C: Affect regulation and pain in borderline personality disorder: a possible link to the understanding of self-injury. Biol Psychiatry 2010;68: 383–391.
35 Pessoa L, McKenna M, Gutierrez E, Ungerleider LG: Neural processing of emotional faces requires attention. Proc Natl Acad Sci U S A 2002;99:11458–11463.
36 Ochsner KN, Ray RD, Cooper JC, Robertson ER, Chopra S, Gabrieli JD, et al: For better or for worse: neural systems supporting the cognitive down- and up-regulation of negative emotion. Neuroimage 2004;23:483–499.
37 Breiter HC, Etcoff NL, Whalen PJ, Kennedy WA, Rauch SL, Buckner RL, et al: Response and habituation of the human amygdala during visual processing of facial expression. Neuron 1996;17:875–887.
38 Niedtfeld I, Kirsch P, Schulze L, Herpertz SC, Bohus M, Schmahl C: Functional connectivity of pain-mediated affect regulation in borderline personality disorder. PLoS One 2012;7:e33293.
39 McRae K, Hughes B, Chopra S, Gabrieli JD, Gross JJ, Ochsner KN: The neural bases of distraction and reappraisal. J Cogn Neurosci 2010;22:248–262.
40 Boettiger CA, D'Esposito M: Frontal networks for learning and executing arbitrary stimulus-response associations. J Neurosci 2005;25:2723–2732.
41 Ochsner KN, Bunge SA, Gross JJ, Gabrieli JD: Rethinking feelings: an FMRI study of the cognitive regulation of emotion. J Cogn Neurosci 2002;14: 1215–1229.
42 Rainville P: Brain mechanisms of pain affect and pain modulation. Curr Opin Neurobiol 2002;12: 195–204.
43 Treede RD, Apkarian AV, Bromm B, Greenspan JD, Lenz FA: Cortical representation of pain: functional characterization of nociceptive areas near the lateral sulcus. Pain 2000;87:113–119.
44 Kluetsch RC, Schmahl C, Niedtfeld I, Densmore M, Calhoun VD, Daniels J, et al: Alterations in default mode network connectivity during pain processing in borderline personality disorder. Arch Gen Psychiatry 2012;69:993–1002.
45 Zanarini MC, Frankenburg FR, Reich DB, Fitzmaurice G, Weinberg I, Gunderson JG: The 10-year course of physically self-destructive acts reported by borderline patients and axis II comparison subjects. Acta Psychiatr Scand 2008;117:177–184.
46 Chapman AL, Gratz KL, Brown MZ: Solving the puzzle of deliberate self-harm: the experiential avoidance model. Behav Res Ther 2006;44:371–394.
47 Kleindienst N, Bohus M, Ludäscher P, Limberger MF, Kuenkele K, Ebner-Priemer UW, et al: Motives for nonsuicidal self-injury among women with borderline personality disorder. J Nerv Ment Dis 2008; 196:230–236.
48 Nock MK, Prinstein MJ: A functional approach to the assessment of self-mutilative behavior. J Consult Clin Psychol 2004;72:885.
49 Reitz S, Krause-Utz A, Pogatzki-Zahn EM, Ebner-Priemer U, Bohus M, Schmahl C: Stress regulation and incision in borderline personality disorder – a pilot study modeling cutting behavior. J Pers Disord 2012;26:605–615.

50 Dedovic K, Renwick R, Mahani NK, Engert V, Lupien SJ, Pruessner JC: The Montreal Imaging Stress Task: using functional imaging to investigate the effects of perceiving and processing psychosocial stress in the human brain. J Psychiatry Neurosci 2005;30:319–325.

51 Reitz S, Klütsch R, Niedtfeld I, Knorz T, Lis S, Paret C, et al: Incision and stress regulation in borderline personality disorder: neurobiological mechanisms of self-injurious behavior. Br J Psychiatry 2015, Epub ahead of print.

52 Fimer I, Klein T, Magerl W, Treede R-D, Zahn PK, Pogatzki-Zahn EM: Modality-specific somatosensory changes in a human surrogate model of postoperative pain. Anesthesiology 2011;115:387–397.

53 Shabes P, Baumgärtner U, Schmahl C, et al: Comparison of different pain modalities with incision pain (abstract). 3rd International Congress on Borderline and Allied Disorders, ESSPD, Rome, 2014.

54 Naoum J, Schmahl C, Reitz SC, et al: The role of seeing blood in non-suicidal self-injury – an experimental approach (abstract). 3rd International Congress on Borderline and Allied Disorders, ESSPD, Rome, 2014.

Prof. Dr. Christian Schmahl
Department of Psychosomatic Medicine and Psychotherapy
Central Institute of Mental Health, Medical Faculty Mannheim, Heidelberg University
DE–68159 Mannheim (Germany)
E-Mail christian.Schmahl@zi-mannheim.de

Author Index

Subject Index